A2 UNIT 2

STUDENT GUIDE

CCEA

Geography

Processes and issues in human geography

Tim Manson

HODDER
EDUCATION
AN HACHETTE UK COMPANY

Orders: please contact Hachette UK Distribution, Hely Hutchinson Centre, Milton Road, Didcot, Oxfordshire, OX11 7HH. Telephone: +44 (0)1235 827827. Email: education@hachette.co.uk. Lines are open from 9 a.m. to 5 p.m., Monday to Friday. You can also order through our website: www.hoddereducation.co.uk.

ISBN 978-1-4718-6410-0

First published in 2017 by
Hodder Education,
An Hachette UK Company
Carmelite House
50 Victoria Embankment
London EC4Y 0DZ
www.hoddereducation.co.uk

First printed 2017

Impression number 6

Year 2023

Cover photo: Alastair Hamill

Typeset by Integra Software Services Pvt Ltd, Pondicherry, India

Printed and bound by CPI Group (UK) Ltd, Croydon, CR0 4YY

Hachette UK's policy is to use papers that are natural, renewable and recyclable products and made from wood grown in well-managed forests and other controlled sources. The logging and manufacturing processes are expected to conform to the environmental regulations of the country of origin.

Contents

Content Guidance

Questions & Answers

■ Getting the most from this book

Exam tips

Advice on key points in the text to help you learn and recall content, avoid pitfalls, and polish your exam technique in order to boost your grade.

Knowledge check

Rapid-fire questions throughout the Content Guidance section to check your understanding.

Knowledge check answers

1 Turn to the back of the book for the Knowledge check answers.

Summaries

■ Each core topic is rounded off by a bullet-list summary for quick-check reference of what you need to know.

■ Option A

Question 1 Cultural geography

Exam-style questions →

(a) Explain why the global growth of cyberspace has been uneven. [8 marks]

ⓔ Marks are awarded for an answer that explains some of the contrasts across the world that deal with economic, social and political issues. Answers will deal with at least two of the three issues.

Level 3 (6–8 marks): Answer addresses at least two of the issues and focuses on 'why' the differences have occurred rather than just a description of the patterns.

Level 2 (3–5 marks): A good quality answer that might only look at one issue or might look at two with less depth.

Level 1 (1–2 marks): Answer is a description rather than an explanation.

Student answer

Sample student answers

Practise the questions, then look at the student answers that follow.

(a) The amount of internet access that people get around the world is varied. People who live in rich countries such as the USA and the UK have high amounts of access. Over 80% of people in the UK have access to the internet. This is because people who live here can afford the expense of having broadband access or 4G access on their mobile phones. It is much easier to get access in the UK as there is more competition for business and competitive prices.

ⓔ 4/8 marks awarded The answer has some detail about the global contrast but it only deals with some of the economic impacts. This restricts the answer to Level 2; the answer would need to address either more social or more political issues to go into Level 3.

Commentary on sample student answers

Read the comments (preceded by the icon ⓔ) showing how many marks each answer would be awarded in the exam and exactly where marks are gained or lost.

(b) Examine the relationship between social inequalities and religion. You should refer to places that you have studied to illustrate your answer. [9 marks]

ⓔ Many people feel that they experience social inequalities such as social exclusion and discrimination based on a range of issues. The answer here will need to refer to social exclusion and discrimination but should also look at how some people's religious beliefs play a role in how they get on with and integrate with others.

Commentary on the questions

Tips on what you need to do to gain full marks, indicated by the icon ⓔ

Level 3 (7–9 marks): There is good commentary on social inequalities (and possibly discrimination) through religion. There is some appropriate reference to places.

Level 2 (4–6 marks): A good quality answer that might only take a superficial look at the social inequalities through religion. Perhaps less detail in relation to places.

Level 1 (1–3 marks): The answer lacks understanding and does not address the key ideas behind the question.

■ About this book

Much of the knowledge and understanding needed for A2 geography builds on what you learned for GCSE and AS geography, but with an added focus on the development of specific knowledge of the impact of human geography. This guide offers advice for the effective revision of **A2 Unit 2: Processes and issues in human geography**, which all students need to complete.

The A2 Unit 2 external exam paper tests your knowledge and application of human geography processes and issues. The exam lasts 1 hour 30 minutes and makes up 24% of the final A-level grade.

To be successful in this unit you have to understand:
- the key ideas of the content
- the nature of the assessment material — by reviewing and practising sample structured questions
- how to achieve a high level of performance in the exam

This guide has two sections:

The **Content Guidance** summarises some of the key information that you need to know to be able to answer the examination questions with a high degree of accuracy and depth. In particular, the meaning of key terms is made clear and some attention is paid to providing details of case study material to help to meet the spatial context requirement of the specification. Students will also benefit from noting the exam tips, which will provide further help in determining how to learn key aspects of the course. Knowledge check questions are designed to help learners to check their depth of knowledge — why not get someone else to ask you these?

The **Questions & Answers** section includes some sample questions similar in style to those you might expect in the exam. The sample student responses to these questions and detailed analysis will give further guidance in relation to what exam markers are looking for to award top marks.

The best way to use this guide is to read through the relevant topic area first before practising the questions. Only refer to the answers and comments after you have attempted the questions.

Content Guidance

■ Option A Cultural geography

Cultural geography

Introduction to cultural geography

Cultural geography encourages us to study how people address some of the issues concerned with their identity. Where does the sense of identity come from? What are the main characteristics of identity and how can people group together into particular cultural groups that share the same characteristics? The themes in cultural geography address issues related to development such as colonialism and post-colonialism, the growth and importance of popular culture and consumption, globalisation, gender and sexuality, race, anti-racism, ethnicity, ideology, language, economics and psychology. It is a celebration of the differences that exist between people. We do not live in a world where everyone is the same (monoculture) but one that is filled with cultural diversity.

> **Exam tip**
>
> Make sure that you have a good understanding of what exactly cultural geography is and know the key terms.

Why do cultural groups exist?

Cultural diversity exists when there are a number of different cultural groups or societies in one place. Some of the more obvious differences between people are language, dress, religious beliefs, morality and traditions.

The United Nations Educational, Scientific and Cultural Organization (UNESCO) noted that 'Cultural diversity is the common heritage of humanity and is a source of exchange, innovation and creativity. Cultural diversity is as necessary for humankind as biodiversity is for nature. Therefore, the cultural heritage of humanity should be recognised and affirmed for the benefit of present and future generations.'

Members of a **cultural group** have a similar outlook. In psychology, Social Identity Theory introduces the idea (from Henri Tajfel) that social identity is 'a person's sense of who they are based on their group membership(s)'. He noted that the sense of cultural identity that a person feels helps to generate a sense of ownership, pride and self-esteem. However, the side-effect of this is that people often generate rose-tinted glasses that will help to increase the self-image of the group so that it develops a superiority over other similar groups.

> A **cultural group** is a community of people who share common experiences.

People often try to band themselves into groups who share their core beliefs and identity. There is a safety and familiarity in being among like-minded people. Distance and lack of contact between groups causes the differences between groups to grow.

Differences between cultural groups

Differences in cultural groups can refer to race, national origin, gender, class or religion. Less obvious differences such as changes to economic status, migration or disability can also bring people together.

> **Knowledge check 1**
>
> What is a cultural group?

There are similarities with the definition of what makes up an **ethnic group**. Membership of an ethnic group is usually characterised by a shared sense of heritage, origin, religion, art, physical appearance or ancestry. Ethnic groups are usually minority groups within the wider social and cultural context.

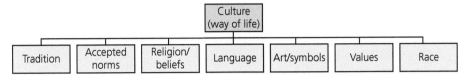

Figure 1 Main aspects of cultural diversity

The building blocks of race, language and religion often help to define the particular group characteristics. History and tradition play an important part in embedding these differences and in passing them on from one generation to the next. This then creates a set of unofficial 'rules' or 'accepted norms' within the society. These are articulated in the values adopted by the group and expressed using different methods of art and symbols that clearly show what separates the group from society at large.

Expression of cultural nationalism

The national identity of a place will be shaped by particular shared cultural traditions and language but *not* on ideas of a shared common ancestry or race. It incorporates (and encourages) feelings of cultural pride that people might have for an area. **Cultural nationalism** allows people who were not born in a place to gain a connection with the place. Migrants can adopt the culture of their host country; they can follow the cultural traditions and holidays, accept the political ideologies, develop the same taste in food or learn the language used by the majority population. Someone can move from Turkey to the UK and can adopt similar attitudes towards dress and food and can learn to speak fluent English. The USA is a good example of successful cultural nationalism. The scale of diversity in the USA is on a different level to many other countries. The USA has managed to absorb a wide number of different ethnicities, yet each migrant has learned English and adopted American cultures and customs (like Thanksgiving).

Social inequalities

Social inequalities occur when resources are not distributed equally among the people. There will be an element of discrimination where some people are seen as being more 'deserving' of help/money/aid/jobs while others are seen as being 'undeserving'. Most places claim to have adopted 'meritocracy' principles where any resources will be distributed on the basis of merit.

In the UN Universal Declaration of Human Rights (UNDHR), Article 2 notes: 'Everyone is entitled to all the rights and freedoms set forth in this declaration, without distinction of any kind, such as race, colour, sex, language, religion, political or other opinion, national or social origin, property, birth or other status.'

Some groups in society will feel that the general society that they live in has not allowed them the same rights that other people have been able to get. They will feel that their liberty and identity is compromised as a result of this.

An **ethnic group** is where a group of people are part of a wider population and have a distinct culture of their own where the group members feel a common origin/identity, are set apart.

Cultural nationalism is a form of national sentiment/feeling/ identity where the nation is defined by having a shared culture.

Exam tip

It is important that you have a clear understanding of what cultural nationalism is. How is this different from ethnic nationalism?

Knowledge check 2

What is cultural nationalism?

- **Social exclusion**: when a group of people are disadvantaged socially and left on the fringes of society. Individuals or groups might be blocked from the rights/opportunities/resources that are available to other people. Individuals in a particular group are denied resources that are fundamental to social integration such as housing, jobs, access to health care or civil rights (e.g. the right to vote/right to marry).
- **Discrimination**: when one person or group of people are treated differently from another person or group of people based on their class or the cultural/social group to which they belong. It is usually when some people are at a disadvantage compared to the rest of society and involves some form of exclusion or rejection.

There are a number of ways in which people within social groups can experience specific examples of social exclusion and discrimination.

Ethnicity

The Institute of Race Relations in the UK describes **ethnicity**/ethnic groups as 'a group of people whose members identify with each other through a common heritage, often consisting of a common language, common culture (which can include a religion) and/or an ideology which stresses a common ancestry. It is the way that most countries and peoples choose to delineate groups.'

One of the key aspects that unites people is the shared experience of coming from the same place. Many people have a strong link with their country of origin and associate with the outward signs and symbols connected with this place, such as national flags, anthems, customs and traditions. Nationality can be both unifying and divisive — when people feel part of the 'group' they have a sense of belonging, but when people do not feel part of the 'group' this can lead to a feeling of discrimination and can lead to conflict.

Most people connect their sense of nationality with the membership of a particular ethnic group that is grounded in a particular place or country. Nationality is usually identified as being part of the population of a nation-state, but sometimes nationality might be expressed as a hope to create a nation-state. For example, many of the Kurdish people in the Middle East identify themselves as Kurds, even if they live in Iraq or Syria.

Ethnic groups will usually proclaim their shared identity and national unity to the outside world. In Northern Ireland (NI) the usually Catholic nationalist population might see their national identity tied more with the Republic of Ireland than with the UK, whereas the usually Protestant unionist community might see their national identity as British and tied with the United Kingdom (UK).

Gender

In many countries the status of women is not the same as that of men; women remain an underprivileged group. Although there have been many advances in the equality of women across the world, there are still many countries where women are treated as second class citizens and where they do not have the same rights or get the same pay as their male counterparts. In many Muslim countries, women have to adhere to strict

Exam tip

You need to be able to look at how social exclusion and discrimination help to explain the links between social inequalities and list of other factors such as ethnicity, gender, race, religion, sexuality and social class. Be aware that questions might focus on only one of these factors.

Ethnicity refers to the process of belonging to or identifying with a particular ethnic group. It is all about the things that unite people in a particular group and what makes the people different from the rest of the population.

rules about what they wear, who they speak to, how they walk behind their husbands and whether they can do individual tasks such as driving.

In 1978 the UN adopted the Convention on the Elimination of Discrimination against Women which was intended as a bill of rights for women. Some of the countries that have not ratified the convention are Iran, Palau, Somalia, Sudan, Tonga and the USA.

The USA is one country where equal pay does not exist. Research shows that the average female salary is usually around 80% of that of the average male salary.

The UN convention aims at making sure that there is gender equality in each country. In recent years the main focus has been on making sure that women can escape the burden of getting involved in forced marriages or in human trafficking (usually for the purposes of sexual exploitation).

Race

The study of **race** and the discussion about the development of the various taxonomies that describe the differences between races is something that will continue for many years. Traditionally, groups of human beings were identified in biological anthropology in three distinct groups.

Race is a categorisation of humans based on the differences in their physical traits, features, ancestry, genetics or social relations.

- **Caucasian** (or Caucasoid or Europid): this usually refers to the areas from which the people had come but later also related to a light skin tone and particular physical characteristics such as cranial features, nose structure and hair colour and type.
- **Mongoloid**: this has been used to describe people from different areas of Asia. The vast majority are described as having straight black hair, dark brown almond-shaped eyes and many have broad, relatively flat faces.
- **Negroid** (or Congoid): this usually defines people from sub-Saharan Africa but has also been expanded to include Australoid native peoples from the South West Pacific including Australia and New Zealand. Again, differences in skull shape, nose structure, narrower ears, narrower joints and differences in hair and skin pigmentation are evident.

The classification and use of race to identify and separate people has been a cause of division and conflict for centuries. Many people around the world have been affected by racial discrimination. Black people living in tropical climates were seen to need to be 'civilised' by their white superiors. The discrimination allowed slavery to develop to the extent that many West African countries struggled to cope with the loss of so many people (especially men). Slavery was a cause of the American Civil War, yet many black Americans were unable to vote until the 1960s and had to fight for their civil rights.

Religion

The formality of some world religions can give followers a distinct sense of 'being different'. Membership of a particular group can give people a particular mode of behaviour, pattern of daily life, ways to dress, eat and treat family members. However, it is often the individual code of behaviour that can identify someone as being part of a particular religious group. Orthodox Jews and Muslims have particular dress codes that make them stand out from the rest of the population. Jewish people might have a tallit (prayer shawl), kippah (skull cap), tichel or snood (head scarf for women). Muslim women might wear a hijab/khimar (headscarf) or burqa (full-face veil).

Many world conflicts have involved aspects of religious discrimination. In 2014, religious hostilities increased in every area of the world apart from the Americas. That same year, religious-related terrorism and sectarian violence took place in around 30% of all of the Middle Eastern and North African countries. The Christian group Open Doors noted that 2,123 Christians were killed for their faith in 2012. Wars in Sudan, Ethiopia and Somalia have religious undertones, while potential unrest in Nigeria, Afghanistan, India, Pakistan, Palestine, Egypt and Indonesia all can be traced back to issues between religious groups.

Sexuality

In many countries there are specific laws that mean that people who have different sexual attitudes might experience some degree of social exclusion or discrimination. In 2016, it was still a criminal offence for same-sex sexual contact in 74 countries (many of these are across Africa and the Middle East). In 13 countries being gay or bisexual is punishable by death: Sudan, Iran, Saudi Arabia, Yemen, Mauritania, Afghanistan, Pakistan, Qatar, UAE, Nigeria, Somalia, Syria and Iraq.

The Northern Ireland Equality Act 2006 was government legislation to outlaw discrimination in the provision of goods, facilities, services, education and public functions due to sexual orientation.

Although the Republic of Ireland was the first country in the world to legalise same-sex marriage by popular vote, in NI in 2015, one of the major political debates involved whether to legalise same-sex marriage. An NI Assembly vote in November 2015 gained 53 votes for and 52 votes against. However, the motion was blocked by the DUP when they applied a petition of concern that required the proposal to achieve a cross-community majority.

Sexualism is when discrimination takes place based on an individual's sexuality/ sexual orientation or sexual behaviour. Usually this is when heterosexual people are biased towards lesbian, gay, bisexual or asexual people.

Social class

The social class that a person has within society will impact on the perception that they have of themselves and the perception that others have of them. These roles can come from many different sources; there might be influence from social roles within the family, or they could be associated with roles within the world of work or politics.

In the past, divisions in social class, status or even caste have created ethnic and social groups. Many classifications of social class are based on economics. The amount of money that someone earns will often directly influence the class to which they are perceived to belong.

Table 1 Socio-economic classification for the UK (ONS, 2001)

Group	Description
1	Higher professional and managerial workers
2	Lower managerial and professional workers
3	Intermediate occupations
4	Small employers and non-professional self-employed
5	Lower supervisory and technical
6	Semi-routine occupations
7	Routine occupations
8	Long-term unemployed

Today in the USA, the social class distribution is directly linked to income. The 'American Dream' is an ideal that opportunity is available to any American, allowing the highest aspirations and goals to be achieved. This means that someone can be born into poverty but still has the opportunity to climb the social ladder to the very top.

Social constructions of nature and landscape

A social constructionist approach is usually seen as being the perspective that many of the characteristics and groups that shape society are inventions made up by the people living within them. It is a method of organising things into a hierarchy. Geography as a subject is a social construction, a product of the interactions of many different thinkers over a long period of time. For those of us who describe themselves as 'geographers', this makes Geography 'our' subject; we are the ones who use it to help construct our attitudes and outlook towards things like nature and the landscape. It is often the people who hold an idea of social construction who can therefore start to challenge and shape it further.

Social construction suggests that things usually historically associated with the human body, such as gender and race, are constructed socially or invented rather than through some biological study. Traits associated with race might be seen as 'low intelligence' or 'uncivilised' behaviours; these will come from perceptions that people have rather than their physical makeup.

Landscapes as human systems

Geographers often see landscapes as more than just the physical features that make up and shape the land. Landscapes are the symbolic environments created through human acts so that particular meanings are transferred onto the environment and give the environment a new definition. Greider and Garkovish (1994) note that 'these landscapes reflect our self-definitions that are grounded in culture'.

Landscapes change over time and so too will the attitudes that make up a social construction. Once ideas have been constructed, they can become independent of their constructors. Landscapes therefore have become the symbolic environments that are created by humans as they try to add meaning to nature and the environment. People with different viewpoints will look at nature in different ways; when looking at an empty field an estate agent will see opportunity, a boy will see a football pitch, his mum might see the need for a separate wash in the washing machine, a farmer might see crops for harvesting, an environmentalist might see animal habitats that need protecting. How we look at a landscape will reflect our self-definition which will be rooted in the culture that we belong to.

Natural landscapes

Some geographers would argue that a social constructionist approach does not allow an identification of anything being 'natural'. Each individual person will have a different understanding of nature.

Films like *Jurassic Park* remind us of the countless ways that people try to influence (and control) nature. Nature, and the landscapes that it forms, can be a social

Exam tip

The discussion here about social construction is not straightforward. Make sure that you are clear about the differences. What are social constructions? How will this affect how people look at different aspects of the landscape?

creation. It is constantly being changed and remade. It is adapted, fought over and owned by people. When we see greenery, we do not necessarily see the vegetation or reflect on the climatic factors that have allowed the plants to grow. Instead, we might think this is a potato plant, planted by a farmer, on his land, surrounded by hedgerows that have divided up this land for 250 years. Raymond Williams remarked, 'the idea of nature contains, though often unnoticed, an extraordinary amount of human history'.

Worldwide there are many examples where the landscape is defined and viewed differently by the cultural group that lives there, e.g. in India the World Bank has helped to develop a series of dams and irrigation projects along the Narmada River. On the one side these projects are required to help prevent flooding and to provide much-needed water to farmland. However, the river is also a Hindu holy river and is revered for its healing properties. Each of these viewpoints is a valid experience and understanding of the importance of the river and so informs any discussion about the river.

Cultural landscapes

Social constructions can also have an impact on the cultural breakdown or differences across a particular landscape. In every society, people will have constructed a mind-set of social reality that they have implemented into a set of social or cultural practices.

For example, social exclusion and discrimination can be adopted as an important part of a cultural landscape. People who have adopted ideas of racial superiority can have a major impact on the politics, laws and policing of a country. This can also flavour the value that people put on human life, the rights that native people might have for land ownership in the face of colonial rule, or might restrict the rights of women within society.

There are cultural differences in how people will construct their concept of themselves within the social context. An individual needs to play a part within a wider group context. The group dynamic and experience will dictate how it interprets its place within the cultural landscape.

Summary

- Cultural geography encourages us to study how people address some of the issues concerned with their identity.
- Cultural diversity is when there are a number of different cultural groups in one place who share similar experiences and have a similar outlook.
- Cultural nationalism is the national sentiment that people feel when they have a shared cultural identity. It allows people not born in a place to adopt the culture of their host.
- Social inequalities exist within society and cause levels of social exclusion and discrimination that are linked to ethnicity, gender, race, religion, sexuality and social class.
- There are a number of social constructions created by people to help interpret nature and the landscape.
- Landscapes can be seen as human systems and as natural and cultural landscapes.

Migration

Every day people are on the move from one place to another and the reasons why people feel the need to move are as varied as the places where they come from.

Mobility is the movement of people from one place to another. Usually this can involve a circulation movement or **migration**.

Migration can also take place internally, within a country; 'in-migration' and 'out-migration' are used to describe when people move from one part of a country to another (e.g. students travelling to Scottish universities from NI).

Push/pull factors in migration

Push factors are seen as the things that cause people to want to leave. These are the things that make them unhappy with their current life. **Pull** factors are the things that are attractive in the new place, things that might be seen as being improved due to a move.

The decision that people take about whether to migrate or not can be based on a number of factors.

Economic factors

It is economic pressures, concerns or opportunities that provide the main stimulus for a migration move. People want to improve their wages and will be prepared to move to facilitate this.

The majority of people who migrate are doing so in order to try to improve themselves. An economic migrant is someone who wants to move to get a job, to earn money or to improve the amount of money that they have been earning to date. Often people who are prepared to migrate will take jobs that do not use the skills in which they have been trained in their home country.

A good example of this followed Poland's accession to the EU in 2005 which led to a lot of emigration from Poland to NI and England. Over the next eight years there was a marked increase in the number of people moving so that they could get better-paid jobs elsewhere in Europe.

Figure 2 shows the estimate of international migration into NI from 2001 to 2007 and, in particular, the impact of the A8 countries joining the EU.

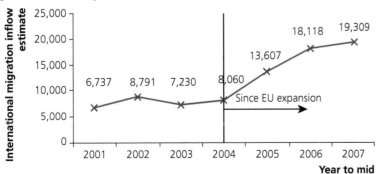

Figure 2 Estimate of long-term international in-migration to NI (2001–2007)

Migration is the geographic movement of people across a specified boundary with the aim of establishing a new permanent or semi-permanent residence. Migration is one of the main components of population change (along with fertility and mortality).

Immigration refers to a movement of people between countries (international migration) when people are moving into a country. **Emigration** is when people are leaving a country.

Gross migration looks at the volume of migration in an area and includes all of the migration flows into and out of an area. **Net migration** is the balance of migration and is the difference between the in and out flows for an area.

Knowledge check 3

What is the difference between push and pull factors of migration?

Social factors

Another reason for migration is to join others who are in the same family who have maybe made an earlier move. Many migratory moves started with the main bread-winner (usually male) moving and then once they have been established, they are followed by the wife and family.

Students might move to a city for university (for the social life) and many will stay on when their studies are over. Sometimes the availability of housing, health care and education facilities will tempt a family to choose one place over another.

Political factors

Often the reasons why people move are influenced by the political landscape. They will move away from potential victimisation or threats from political enemies. There might be conflict or war in the area. Whether someone can actually migrate into another country or not will be decided by the level of control on immigration set by the country and the policies set out by the government. Excessive amounts of immigration into a country do not go down well with the host population and often controls are required on the level of migration.

Cultural factors

People who do not have religious or personal liberty to do what they want or to worship in the way that they want will often make a decision to move elsewhere. The Pilgrim Fathers who migrated to America faced some persecution. Some of the migrants who moved to Ulster as part of the Ulster plantation would have moved for similar reasons.

Today many sections of large cities are inhabited by groups of ethnicities from different countries. These enclaves provide an element of security and familiarity.

Environmental factors

Environmental crises such as droughts, earthquakes, floods, landslides, desertification and environmental accidents can all have a big impact on migration. Disasters like Chernobyl caused the evacuation of more than 100,000 people. More recently, over 300,000 people have been evacuated following the exposure to radiation caused at the Fukushima nuclear reactor in the aftermath of the tsunami triggered by the Tohoku earthquake on 11 March 2011.

Barriers to migration

The decision to migrate is never taken lightly. The migrant has to balance the pros and cons of moving and then decide whether to go ahead with the move or not. A series of barriers might get in the way and influence this decision; maybe a family member becomes sick, transportation costs increase or financial difficulties make it hard to sell a home before leaving.

In addition, the rules for migration between countries are often such that migratory moves take a long time to process and by the time the government gives permission to move, a job opportunity might have been lost.

E.S. Lee (1966) used a variety of new hypotheses as a framework for the investigation of the spatial, temporal and causal factors in migration. He reinstated the basic push–pull concept instead of isolating the pressures and stimuli that confronted particular individuals and groups.

Lee suggested that every individual is exposed to a set of positive and negative factors which govern the move. Lee argues that these will vary according to age, sex, education or marital status and that decisions can be modified by the intervening obstacles, such as legal restrictions, family attachments, personal anxiety, costs of the move etc.

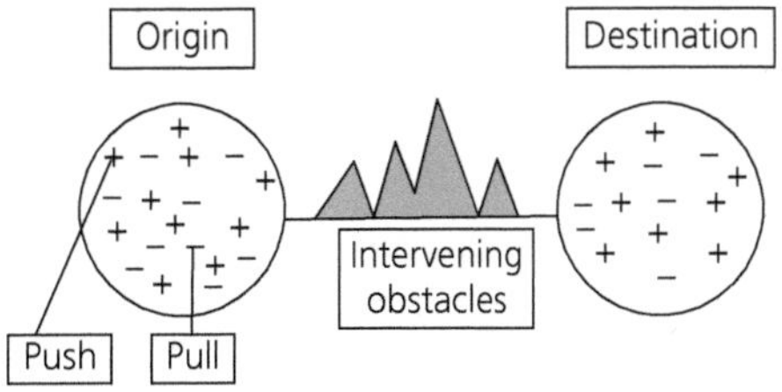

Figure 3 The intervening obstacles to migration

Implications of migration

Service provision

As people move into an area there can be a large amount of pressure put on resources in that area. Basic services such as water, electricity and sewerage can come under stress when too many people move at once, e.g. in slums/shanty towns. However, more general services such as schools, hospitals, doctors, shops, housing, transport and access to jobs can also cause services to reach breaking point.

Equally, as people move away from an area, the decrease in demand can result in services being lost or consolidated towards larger centres of population.

Economic activity

The direct intention of most migrants is to earn more money than they would at home so that they can send money home either to improve the standard of living for their family or to save enough to bring them over to the new country. It is important that the economic migrant makes money and saves money. Remittances are the amounts of money that migrant workers send back to their families. In some countries, these remittances are a vital source of income, often accounting for 10–15% of the GDP of countries like Turkey.

Social stability

Any mass movement of people to or away from an area can have big consequences for the social stability and the fabric of society. Huge changes can be brought by an influx of people as food, language, religious observances and dress can all change to highlight a more cosmopolitan society. This can bring tensions as some natives feel threatened and want to remove the new arrivals.

Equally, young migrants leaving an area can cause the area to be populated solely by elderly people who might not be able to keep doing the same work. Services will fall apart quickly and decisions will have to be taken to abandon houses and areas altogether.

Knowledge check 4

What are the main barriers to migration?

Exam tip

Make sure you can explain the way that push and pull factors work through each of the aspects of migration.

Exam tip

You should have a general understanding of the different impacts/implications that migration can have on service provision, economic activity and social stability before you work through the case studies.

Case study

Implications of out-migration on a small scale

The Blasket Islands are a small group of islands located just off the southwest coast of the Dingle peninsula in the Republic of Ireland. The decline of this rural community over time (Table 2) led to the eventual evacuation of the islands in 1953. Many of the descendants went to the USA and a migration stream linked them to Springfield, Massachusetts. The islands remain uninhabited today.

Economic activity

The island people who lived on Great Blasket Island lived traditional subsistence/crofting lifestyles where each family would survive by trapping rabbits, poaching eggs, fishing and doing some pastoral and arable farming. There was never a surplus of cash from jobs. The workforce were relatively unskilled, but had to learn the skills that allowed the population to survive living on this rock in the Atlantic. At its peak there were 175 residents on the islands.

From 1845 to 1851 the Great Famine decimated the population of Ireland. About one million people died in the famine and another million emigrated to England or to the USA. However, the islanders did not suffer as much as the mainlanders.

Many of the islanders received remittances from family members who had migrated overseas. After the Second World War, many of the young people received passage to go to live with their relatives in the USA. As the youth population left, this meant that jobs (such as collecting eggs from nests high up on cliff faces) had to be done by the ageing population.

Table 2 Population of the Blasket Islands

Year	Population	Year	Population
1821	128	1938	106
1841	153	1947	50
1851	97	1953	22
1930	121		

Service provision

There were few services on the island to begin with. There were no formal shops, post offices, pubs or religious services. A national primary school was established in 1860 but this closed in 1940. At its height there were 50 children at the school but this declined to 4 in 1940. A telephone and small post office arrived in 1930. There was no electricity, no sewerage, no mains water and poor transport links with the mainland. There was rarely much money on the islands to pay for services.

As society in Ireland modernised, the people who lived on the island, the young people in particular, felt increasingly isolated and left behind. The young people saw no future on the island. Fishing became increasingly insecure and life elsewhere became more attractive.

By 1953, the sustainability of being able to provide more modern lifestyles for the rural population of Ireland was being questioned. The evacuation order by Taoisech (Prime Minister) Eamon de Valera was accepted as a necessity. Only 22 people were left, with little communication and no emergency assistance. There were not even enough men to be able to crew the traditional Naomhog boats that would connect them to the mainland.

Social stability

The islanders were a tight community. They depended on each other and as the population declined the community could not be sustained as it once was. Young men left the islands to find partners but increasingly found it difficult to convince their mainland wives to move back to the island. Young people heard about opportunities and lifestyles in the USA and wanted those for themselves. Gradually the community decreased and people got lonelier as the young refused to return. Diseases spread rapidly and many children died from simple illnesses because they could not get medical care quickly enough.

After the Second World War, opportunities for emigration and employment off the island made the remaining community structure vulnerable. Although remittances kept money coming into the island, these did not compensate for the lack of workers on the land.

Implications of urban in-migration

Following the Second World War, there was a huge labour shortage in Germany. The German economy needed more people to come and work in factories to help manufacture products that would rebuild the economy. The government began to look for labour outside the country and recruitment agreements (*Anwebeabkommen*) were agreed with Italy in 1955, Spain in 1960, Turkey in 1961 and Yugoslavia in 1968.

The city of Munich is located in the area of Bavaria in the south of Germany. It has a population of around 1.5 million people. The city is home to many major companies including BMW, Siemens, MAN, Linde, Allianze and MunichRE. However, over the years the impact of large volumes of migration has continued to affect the city. Munich is one of the top-ranked destinations for migration (Table 3) and hosts 530,000 people with an international background (38% of the entire population).

Table 3 Largest groups of foreign residents in Munich (2013)

Turkey	39,857
Croatia	26,070
Greece	25,574
Italy	24,337
Austria	21,579
Poland	20,103
Bosnia and Herzegovina	15,836
Romania	14,293
Serbia	13,631
Iraq	10,394

Economic activity

Manufacturers in Munich needed a huge amount of labour. Countries like Turkey provided a cheap, skilled, constant supply of eager workers who could fill the labour gap. In total 2.8 million Turks now live in Germany and nearly 40,000 of these live in Munich. The use of additional labour meant that the German economy continued to grow to the point that it was the third biggest economy in the world. Working conditions and wages were the best in the world. However, despite the influx of workers from other

countries, there were still Germans who remained unemployed. Recessions in 1967 and 1990 led to many of the 'white-collar' Germans losing their jobs while the 'blue-collar' Turkish workers retained theirs.

With such a large immigrant population working in the city, some of the money made by these workers was 'leaking' out of the country. They would send remittances back to their families which meant that money was not being reinvested into the local economy.

Service provision

Immigrants into Munich took on the jobs that the Germans did not want to do. These were low-paid jobs at the bottom of the social ladder. Many Turks were employed driving trams and buses or cleaning offices and factories. The migrants helped to provide the services that the German population had become accustomed to and new services were introduced such as Turkish baths, kebabs and Turkish barbers.

However, the influx of so many people (increasing the city's population from 824,000 in 1950 to 1.5 million in 2015) meant that there was a lot of pressure on services. Schools had to cope with students who might not speak German. Migrants might have special requirements for meal times (halal foods) or religious observances (prayer times). Hospital staff found it increasingly difficult to communicate with patients who might speak different languages. An increased number of translation services were required which all cost money.

The European migrant crisis in 2015 caused a huge influx of migrants to travel to Munich. German authorities noted that on some days around 13,000 people were arriving in the city. Officials were quoted as saying, 'We have reached the upper limit of our capacity' as the city workers frantically tried to find accommodation for the new arrivals.

Social stability

Migrants who moved to Germany over the last 50 years brought their own ethnicity and culture with them. They continued to converse in their own languages and observe their religions. Their cultural values were passed on to their children so these traditions would continue. The cultural diversity throughout Germany increased.

The rise of neo-Nazi groups was at odds with feelings that Germany should become a more integrated and multicultural society. A number of violent attacks have taken place (against Turks in particular); these have included a series of arson attacks, bombings and a shooting.

The European migrant crisis in 2015 put further pressure on the stability of relationships between migrants and natives. Up to 1 million refugees and asylum seekers were given entry into Germany. This has caused a huge amount of social and political conflict as native Germans, and those who have been in the country a long time, are worried about security and the pressures on resources that these people might bring to the country.

Migration processes

What is a migration stream?

Migration movement often encourages other people to move from the same source to the same destination. Following the Great Famine in Ireland from 1845 to 1851 many people migrated from rural to urban Ireland, then migrated from Irish cities to English and Scottish cities. Some even migrated internationally to Canada, USA, Australia and New Zealand.

Modern migration streams within Europe show movements of eastern Europeans into the UK and people from Afghanistan and North Africa.

Voluntary and forced migration

Table 4 Voluntary and forced migration

Voluntary migration	Forced migration
■ Economic — jobs: miners moving to South Africa for gold/ diamond mines; Mexicans into the USA	■ Environmental: accidents/disasters, e.g. Fukushima nuclear plant (Japan)
■ Economic — higher salaries: UK doctors and nurses to the USA	■ Resettlements: Ameri-Indians in the Brazilian rainforest
■ Economic — tax avoidance: celebrities to Dublin, Caribbean islands	■ Overpopulation: Chinese people into new cities
■ Social: education and health facilities better in the UK than France/Spain	■ Redevelopment: clearance of poor areas in UK cities
■ Social: welfare state and benefits better in the UK than some other EU countries	■ Natural disasters: earthquakes (Haiti, 2010), volcanoes, mud slides, floods
■ Political: links with colonies across the world	■ Famine: Ethiopia and Sudan in the 1980s and 1990s
■ Cultural — religion: Jewish people deciding to move to USA or back to Israel	■ War: Darfur resettlement camp, Afghans moving to the UK
	■ Religion: Jewish persecution, religious persecution between Muslims and Christians in Nigeria
	■ Slavery: slave trade in the 1600s or the sex trade today

Voluntary migration occurs when people are moving to try to improve their lives. They have a choice as to whether they stay or move. If people have no choice in their migration move, this is **forced migration** and this might be due to natural disasters or some other social, economic or political pressure.

In 2004, the Berne Initiative was a process carried out to try to manage migration better. This would allow governments around the world to agree to the International Agenda for Migration Management which would enable the movement of people in a humane and orderly way. This stated that governments have the responsibility to determine the conditions for entry and stay of all non-nationals. Visas can be used to regulate migration flows in and out of countries. Border control is used to control the movement of people in and out of territories.

Undocumented migrants

Undocumented migrants are people who do not have a permit of residence allowing them to stay in their country of destination. These are people who have not got a visa or permission to move into a country. They might have unsuccessfully sought asylum, overstayed their visa or have entered the country illegally. It is estimated that there are 5–8 million undocumented migrants in Europe.

Illegal migrants

The majority of migrants are prepared to fulfil all of the legal requirements for entry into a country. However, sometimes people will have the intention to enter and remain in a country illegally. This is evident in the UK at present as many people from North Africa travel across Europe to the French port of Calais and then try to get to England on lorries and ferries so they can try to get some work in England.

Asylum seekers

An **asylum seeker** is a person who has applied for asylum under the 1951 Refugee Convention on the Status of Refugees on the grounds that if they returned to their country of origin they would have a well-founded fear of persecution — due to race, religion, nationality, political belief or social group. That person remains an asylum seeker for as long as their application or appeal against refusal of their application is pending.

Refugees

In its broadest context a **refugee** means a person fleeing from things such as civil war or a natural disaster but not necessarily fearing persecution.

People trafficking and migrant smuggling

There has been a recent rise in the number of people who have been victim of human trafficking/smuggling. The UN Convention against Transnational Organized Crime (UNTOC) supports governments in implementing a series of measures that will help to strengthen the relationships required for authorities to work together to make sure that victims are supported and that the trade of people is stamped out.

Immigration as a political issue

Documented and undocumented migration

The 2016 vote in the UK in relation to EU membership was dominated by fears of uncontrolled migration and many voters wanted to ensure that the UK would regain control over its borders. The free movement of European citizens (documented migration) especially from eastern European countries into the UK brought this

---

Knowledge check 5

What are the main differences between documented and undocumented migrants?

Asylum seekers are people who have arrived in a new country but do not have the correct legal requirements for entry. They will seek permission to stay in a country because they will claim that if they have to return home they will face some form of punishment, torture or death.

The United Nations (UN) notes that a **refugee** is 'a person who cannot return to his or her own country because of a well-founded fear of persecution for reasons of race, religion, nationality, political association or social grouping'.

Exam tip

There are a lot of subtle differences between definitions of different types of documented and undocumented migrants. Make sure you are aware of these.

issue to the fore. Political leaders come under pressure to respond to the perceived pressures that excessive migration can bring. Residents felt that the unprecedented levels of migration were putting their lives, services and living spaces at risk. Politicians regularly repeat the issues involved with 'invasion', alleged threats and actual prejudices, which all ensure there is a negative image of migration: concerns about crime, disease, terrorism, detention and surveillance.

The number of people who want to migrate from where they live to another country is at an all-time high. In 2016, one major pressure on the EU was the number of people who attempted to cross the Mediterranean Sea to get to Europe. EU states received over 1.2 million first-time asylum applications. By November over 4,271 migrants had died in the process. In January and February 2016, over 123,000 migrants landed in Greece. Between March 2016 and October 2016 over 140,000 migrants arrived by sea into Italy. The vast majority came from Syria, Kosovo, Afghanistan, Albania, Iraq, Eritrea, Serbia and Pakistan.

The causes behind the huge amounts of immigration are rarely discussed: global inequality, conflict and human rights abuses.

Governments' responses

In the UK, Migration Watch UK noted that 'immigration is a natural part of an economy and society...the problem is the current scale of immigration, nearly half of it now from the EU, which is simply unsustainable'.

They note that the amount of high net migration into the UK in recent years (Figure 4) has caused a rapid population growth. With the UK population standing at 65 million, there are some estimates that it will continue to expand by around 500,000 people each year. This would result in the population increasing by 8 million over the next 15 years. One of the major social issues with this is in making sure that the people integrate into British society.

> ### Knowledge check 6
>
> Why has migration become such a major political issue in recent years?

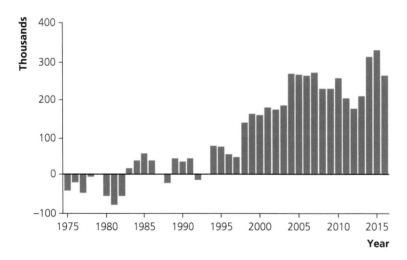

Figure 4 Net migration to the UK (1975–2016)

Case study

Government response to immigration on a national or international scale

Germany has had a long history of immigration. Since the end of the Second World War, many people have immigrated into Germany in order to keep the economy growing (Table 5). Germany is the second most popular destination in the world (after the USA). In the 1960s a shortage of labour during the *Wirtschaftswunder* (economic miracle) led to sizeable numbers of guest workers (*Gastarbeiter*) being brought to live and work in Germany.

Table 5 Top 10 nationalities within the German population (82 million)

German	65 million	Serbian	450,000
Russian	3.5 million	Greek	395,000
Polish	2.8 million	Syrian	366,000
Turkish	2.8 million	Dutch	350,000
Italian	830,000	Austrian	345,620
Romanian	657,000		

Germany has long had a reputation as a country that was prepared to take refugees and asylum seekers. On 1 January 2005 a new immigration law came into force in Germany. For the first time Germany started to see itself as an immigration country. The policy was introduced as the country realised that its ageing population with a declining death rate would soon mean that there would not be enough workers to sustain the retiring population. Foreigners from non-EU countries would be denied access to simple jobs but there was also a raft of opportunities through special regulations like temporary contracts for seasonal workers. People who were deemed as highly qualified would be able to stay permanently instead of leaving after five years.

The European immigration crisis since 2014 has meant that the number of people looking for asylum in Germany has increased hugely. In 2015, Germany took in an estimated 800,000 to 1 million asylum seekers (four times the number in the year before). In 2016, the country welcomed an additional 280,000 asylum seekers.

The German response to the EU migrant and refugee crisis was seen by many as a unilateral open-door policy and it triggered both a domestic and international backlash. The government used a quota system to distribute the asylum seekers throughout the German states.

Some commentators have noted that the German Chancellor Angela Merkel's open-door policy towards migration, in particular towards people fleeing from the Middle East areas of conflict, will make Germany a safer place from terrorist attacks. On 4 September 2015, Merkel said, 'We decided to fulfil our humanitarian obligations. I did not say it would be easy . . . we can manage our historic task — and this is a historic test in times of globalisation . . . Germany is a strong country.' By showing compassion to hundreds of thousands of Muslim refugees, the German people are stating their message that Germany is not at war with Islam.

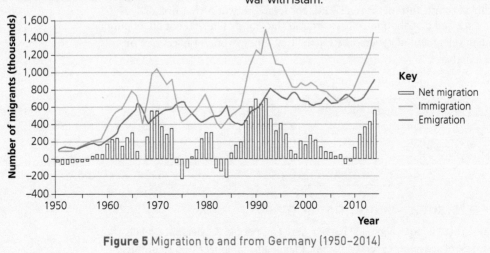

Figure 5 Migration to and from Germany (1950–2014)

Summary

- Migration is usually defined through a series of push and pull factors where people will experience a range of economic, social, political, cultural and environmental factors that will enable them to move subject to some barriers to the migration move.
- A number of implications of migration will affect areas where there is significant out-migration and in-migration. These will have an impact on service provision, economic activity and social stability.
- Voluntary and forced migration will take a number of different forms around the world that create a number of different migration processes and streams including undocumented/illegal migration moves.
- Immigration has become a big political issue (for both documented and undocumented migrants) with differing government responses.

The geographies of cyberspace

The invention of the internet in 1983 was to become one of the most radical inventions in the history of humankind.

Development of cyberspace

In the 1980s Tim Berners-Lee led a team of researchers at CERN in Switzerland that invented the World Wide Web — a method of linking documents to an information system that could be accessed from any node on the network. Earlier in this topic we have seen how people ebb and flow from one place to another and looked at the various impacts and responses. Yet one of the biggest areas for growth in the last 30 years has been among a new virtual or artificial landscape (Table 6). The writer William Gibson has called this landscape 'cyberspace'. In the 1984 novel *Neuromancer*, cyberspace is a dataspace — a vast world in the wires known as the matrix where large transnational companies could trade information. Two of these spaces — the internet and the intranet — are used to show the different ways that geographers can engage with the cyberspace.

The internet is a collection of computers that are linked to networks across the globe. All of these computers and their networks are linked through common communication protocols. Anyone with a computer, modem and telephone line can link into this network. Intranets are private, corporate networks that link offices, production sites and distribution centres around the world. They do not allow public access. The development of cyberspace means that a new world of users can be connected together.

Table 6 The increase in global internet users (2000–2015)

Year	Internet users (million)	% of world population
2000	414	6.8
2005	1,030	15.8
2010	2,023	29.2
2015	3,185	43.4

Changes to socio-economic activity

Theoretically, there is no cyberspace barrier between people. Cyberspace levels the idea of distance of space and social position. People are no longer tied to physical

Knowledge check 7

Why has cyberspace become such an important area for development in recent years?

Exam tip

This is one of the most modern aspects of geography and is going to keep changing. Try to find out the most up-to-date information about how cyberspace and the internet are changing how we look at and organise the world.

anchors of one place but they are free to explore the world through their virtual lenses. This allows geographers to see how people perceive themselves and how they can balance their analogue world with their digital existence. The internet is as diverse as the websites, links and devices that are used to connect to it.

The most profound impact of cyberspace is not in its impact on information processing but on how it affects social relations.

Foucault describes cyberspace as 'the technology of the self'. It is a device that affects the social construction of identity. It is the place where the self is constructed and the rules of social interaction are built. Cyberspaces have become the social spaces where people meet face to face, but under new definitions of what this means. Social media is constantly changing but meeting spaces such as Facebook, Snapchat, Twitter, Facetime and What'sApp all give the opportunity for people to interact with each other in new ways. You can be in the same room as someone or on the other side of the world.

Economically, the internet allows people a greater deal of flexibility and choice. People have access to a much wider range of products and the competitive market means that sellers must ensure that they have realistic prices or they will be unable to sell their product.

Production of international cultures

In 2005, Thomas Friedman introduced his book, *The World is Flat*. Early in the book, Friedman noted that of the ten forces that have 'flattened' the world, each is linked to the development of cyberspace.

The emergence of the internet has become 'a tool of low-cost global connectivity … a seemingly magical virtual realm where individuals could post their digital content for everyone else to access'. The reality of what Friedman argues is that the global competitive playing field has been levelled. This means that the distinctions between rich and poor or MEDC and LEDC have been reduced.

Table 7 World internet usage (2017)

World region	2017 population (million)	% of world population	Internet users (million)	% penetration of regional population
Asia	4,148	55.2	1,856	44.7
Europe	822	10.9	630	76.7
Latin America/ Caribbean	647	8.6	384	59.4
Africa	1,246	16.6	335	26.9
North America	363	4.8	320	88.1
Middle East	250	3.3	141	56.5
Oceania/ Australia	40	0.6	27	68.0

Figure 6 shows clearly that the dominance of the internet in 2002 favoured the richer MEDCs in the northern hemisphere. It was rare for anyone in the LEDC world to be connected. This helps to reinforce the concept of the global digital divide.

Figure 7 comes from www.gapminder.org where there are a number of variables that can be used to measure the development of countries. There is a strong relationship between income and the number of internet users in a country.

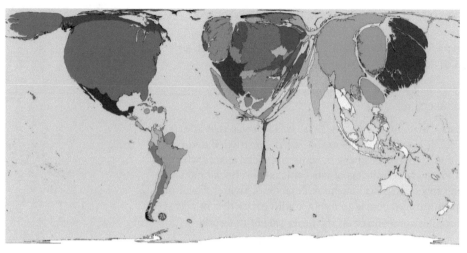

Figure 6 Global map of internet users in 2002 (Copyright Worldmapper.org/ Benjamin D. Hennig)

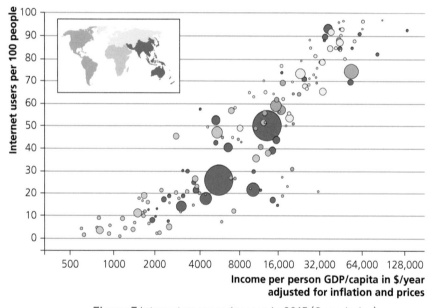

Figure 7 Internet users vs income in 2015 (Gapminder)

Over time, the access that people have to the internet in different countries has been increasing. As a result, people have access to new information and opportunities that they have never had before. One area where there has been a huge growth is in relation to gender. Figure 8 shows the impact that access can have.

Ideas of gender and identity are key to the future of the internet. How will people see themselves? What are their key priorities? Will people who feel that their civil liberties have been compromised in the past be able to use the internet to fuel their path to equality? Does access to the internet allow equality in the world of ideas and access to things? The current rise of the 'internet of things' means that an increasing number of devices have internet connectivity which means that we live in a world that is always online.

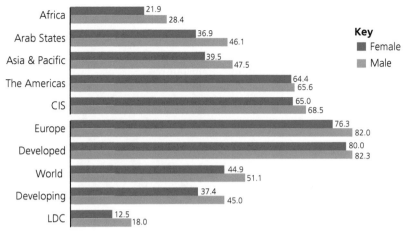

Figure 8 The digital gender gap

Global contrasts

In the early years of the development of cyberspace, the richest places in the world were able to adopt this new technology at a much faster rate. There is a huge contrast in the uptake and use of cyberspace between the global rich and poor.

Economic issues

The world is more connected than it ever has been before.

Bandwidth speed

In the UK there is currently a vast difference in the access that people can get to the internet. Recent reports suggest that parts of rural Wales have the slowest internet across the UK. However, some parts of the world still have no access to mobile phone signal or any form of internet connection. Robert M. Kitchin notes that, 'Globalisation is not an egalitarian process aimed at creating an equitable distribution within and between western countries, and in comparison to developing countries.'

Globalisation of trade

The adoption of the internet as the main source of global trade has allowed more competition and more internationalisation of trading systems. Transnational companies have become increasingly powerful as they can source, manufacture and sell products on a global scale. Costs can be reduced as companies consolidate their overheads by only having one head office, one research and development facility and local manufacturing plants that can get products into the market quickly. For example, the online seller Amazon, founded in 1994, started as an online bookstore but later diversified into selling CDs, DVDs and videos. The founder, Jeff Bezos, bought books from a local wholesaler and sold them online through the internet. Within two months, sales were up to $20,000 per week. Today, the company has revenue of over $135 billion and has more than 341,000 employees who work in a network of 'fulfilment centres'.

Office automation and back office

Much of the work that used to be carried out by office staff can now be carried out using computer technology. Computer programs allow a more efficient approach to planning and financial management. Email allows users to communicate instantly across the world. Many service and call centres can be located at different places around the world to suit the needs of the company, for example mobile phone companies might have call centres in the UK or India.

Teleworking

Teleworking allows workers to use telecommunication services and remain at home, rather than to come into a central office to work. This gives a much more flexible lifestyle to the worker but can also lead to a feeling of isolation and a lack of teamwork ethic.

Employment restructuring

The rise of computer use has meant that there is less need for people to be working in some places. In the USA it was estimated that between 1980 and 1993, the 500 largest transnational companies shed some 4.4 million jobs, most of which were at the professional and technical level. As many of the manufacturing and assembly jobs in the technological industry have been moved to LEDCs, this has meant that the MEDCs keep the 'creative' design jobs but distribute the manufacturing jobs elsewhere.

Social issues

Identity

The rise of virtual reality within the world of the internet allows people to engage with immersive technologies where they can connect with a place they have never been to or they can become different from their normal selves. Foucault suggests that certain technological devices can change the social construction of personal identity. Some people might be more guarded and less likely to speak their mind when online. Others will use the relative anonymity of the internet to engage in personal attacks (trolling) on other people and will express their radical opinions loudly in order to cause offence. Internet users can be who or what they want to be, making it difficult to separate the 'real' and the 'virtual' identity.

Community

The internet has totally changed how we do community. We can now form new types of community based on our interests and affinities rather than the pure coincidence of location. We can generate our own professional learning networks (PLNs). However, there is evidence that most online communities tend to 'retribalise' their new space.

Political issues

Democracy

Through 2016, a series of surprise election results in the UK (Brexit) and in the USA (Donald Trump as President) have helped to emphasise some of the changes that cyberspace has brought to the world of politics. There has been a transformation of political structures, organising, campaigning and voting patterns. Politics is dominated by big party issues and national (or international) events at the expense of more local issues. Cyberspace knows no borders, which means that the laws that govern particular countries get confused and blurred when applied to activities that cross boundaries.

In some countries access to the internet is restricted by governments that might want to make sure that political thought or freedoms are not allowed. In some places like China, North Korea and Saudi Arabia, people do not have full access to the full range of internet services. Governments might lock down different aspects as they want to restrict civil liberties or to disrupt attempts to use the internet to communicate political ideas that are different to those held by the government.

Summary

- The development of cyberspace has brought about some massive changes to socio-economic activity and this has also helped to develop international cultures.

- Global contrasts still remain and these have been brought about by economic, social and political issues.

■ Option B Planning for sustainable settlements

Sustainable development

Explanation of sustainability

What is sustainability?

Sustainability is achieved when the key processes — social (economic, political and cultural) and environmental (ecological) considerations — of **sustainable development** are kept in balance.

The rapid increase of global population continues to put pressure on the economic, social and environmental landscape. On 25 September 2015, the UN adopted a set of goals that were aimed at ending poverty, protecting the planet and ensuring a new prosperity for all. This was part of a new 15-year Agenda for Sustainable Development. The UN General Assembly noted that it was 'resolved to free the human race from the tyranny of poverty and want and to heal and secure our planet. We are determined to take the bold and transformative steps which are urgently needed to shift the world on to a new sustainable and resilient path.'

The 17 sustainable development goals (SDGs, Figure 9), also called the global goals, and 169 targets are unique in that they call for action from all countries while also protecting the planet.

The UN notes that the global goals are aimed at realising human rights for all, achieving gender equality and empowerment within a sustainable development context focused on economic, social and environmental factors.

Goal 11 — **make cities inclusive, safe, resilient and sustainable** — is the main SDG that focuses on sustainability in settlement. The UN is planning a conference on housing and sustainable urban development in Quito, Ecuador in October 2017 (called Habitat III), where world leaders can adopt the New Urban Agenda and help set global standards of sustainable urban development.

Sustainable development is a process to meet the needs of the present without compromising the ability of future generations to meet their own needs. It is meant to create a better life for all people that also maintains a viable future.

Exam tip

Make sure that you have a good understanding of what sustainable development and sustainability are all about.

Figure 9 The global goals for sustainable development

Social considerations

In order to develop sustainability within a settlement, some aspects need to be observed:

- Housing needs to be provided for a range of people and incomes. It should be of high quality and should include health and recreational facilities.
- Poverty and social exclusion need to be addressed, especially in the most deprived areas.
- Improve local environments and areas of industrial decline/brownfield sites.
- Preserve the countryside.
- Health will also be impacted as hospital treatments become increasingly expensive or health concerns are avoided in order to squeeze as many people as possible into an area (e.g. slums/shanty towns).
- Engage further with local people so that local community partnerships are fostered and people take a greater interest in their built environment.

SDG 11 social targets

- By 2030, ensure access for all to adequate, safe and affordable housing and basic services and upgrade slums.
- By 2030, provide access to safe, affordable, accessible and sustainable transport systems for all, improving road safety, notably by expanding public transport, with special attention to the needs of those in vulnerable situations, women, children, persons with disabilities and older persons.
- By 2030, enhance inclusive and sustainable urbanisation and capacity for participatory, integrated and sustainable human settlement planning and management in all countries.
- Strengthen efforts to protect and safeguard the world's cultural and natural heritage.
- By 2030, provide universal access to safe, inclusive and accessible green and public spaces, in particular for women and children, older persons and persons with disabilities.

Knowledge check 8

What are the key points of SDG 11?

- Support positive economic, social and environmental links between urban, **peri-urban** and rural areas by strengthening national and regional development planning.
- Support least developed countries, including through financial and technical assistance, in building sustainable and resilient buildings utilising local materials.

> **Peri-urban** areas describe the land between town and country.

Environmental considerations

In order to develop environmental sustainability within a settlement, some aspects need to be observed:

- Reducing greenhouse gas emissions and improving living conditions including air quality.
- Protecting water resources.
- Ensuring that the management of waste is a priority through the effective use of the waste hierarchy.
- Advancing the use of renewable energy resources.
- Providing a global response to global climate change.

SDG 11 environmental targets

- By 2030, significantly reduce the number of deaths and the number of people affected and substantially decrease the direct economic losses relative to global gross domestic product caused by disasters, including water-related disasters, with a focus on protecting the poor and people in vulnerable situations.
- By 2030, reduce the adverse per capita environmental impact of cities, including by paying special attention to air quality and municipal and other waste management.
- By 2020, substantially increase the number of cities and human settlements adopting and implementing integrated policies and plans towards inclusion, resource efficiency, mitigation and adaptation to climate change, resilience to disasters, and develop and implement, in line with the Sendai Framework for Disaster Risk Reduction 2015–2030, holistic disaster risk management at all levels.

> **Exam tip**
>
> Look at the information about the SDGs on the UN website so you have a deep knowledge of how these might impact on urban areas.

Urban ecological and carbon footprints

Urban ecological footprint

Half of the global population — over 3.5 billion people — live in cities and this number is growing at an excessive rate. The UN notes that '95% of urban expansion in the next decades will take place in developing countries'.

The **ecological footprint** (Figure 10) measures human impact on the Earth's ecosystem. The ecological footprint is the area that a population needs to produce the natural resources it consumes (including plant-based food and fibre products, livestock and fish products, timber and other forest products, space for infrastructure) and to absorb its waste, especially carbon emissions. The ecological footprint tracks the use of six categories of surface areas: crop land, grazing land, fishing grounds, built-up land, forest areas and carbon demand on land.

> **Ecological footprints** measure the demand on and supply of nature.

Urbanisation and migration are causing one of the biggest changes of land use in human history. Cities occupy 3% of land space but account for 60–80% of all energy consumption and make up 75% of carbon emissions.

An urban ecological footprint is the amount of land required to produce the resources needed by one person (to support their lifestyle). It attempts to quantify the impact that one person can make on nature.

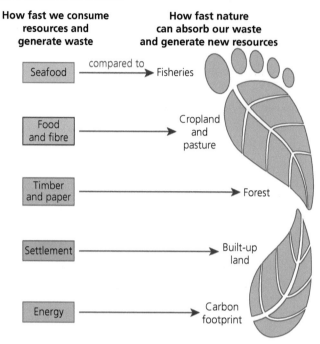

Figure 10 The ecological footprint

Table 8 The five largest urban footprints in England, 2007 (WWF)

City	Planets (needed to support consumption rate)	Footprint (global hectares of land)
Winchester	3.62	6.52
St Albans	3.51	6.31
Chichester	3.49	6.28
Brighton & Hove	3.47	6.25
Canterbury	3.40	6.12

To calculate the footprint, average spending data are used to indicate how resource-intensive the lifestyle of the average citizen is. Spending data are split into eight sectors: housing, transport, food, consumer items, private services, public services, capital investment and other.

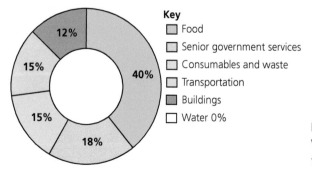

Key
- ☐ Food
- ☐ Senior government services
- ☐ Consumables and waste
- ☐ Transportation
- ☐ Buildings
- ☐ Water 0%

Figure 11 What's in Vancouver's ecological footprint?

Exam tip

It is important that you develop a clear understanding of the subtle differences between urban ecological footprints and carbon footprints.

Carbon footprint

The **carbon footprint** is usually measured by the amount of carbon dioxide and methane emissions within a population.

The supporting theory for carbon footprints is that every action that people take has an environmental consequence. The more resources that people use, the more greenhouse gases will be produced, through the transportation, storage and presentation of products.

There are two types of emissions that make up a carbon footprint — direct and indirect carbon emissions:

- **Direct carbon emissions** come from heating sources and personal transportation.
- **Indirect carbon emissions** come through the production of products: electricity, goods and services and the use of transportation to get goods to market.

How sustainability is related to waste management, energy consumption and water supply

Urban sustainability is based on the idea that cities and urban areas should ensure that they have the smallest environmental impact (ecological/carbon footprint). The modern city should not pollute too much, not consume too many natural resources, allow people space to enjoy their community and make city-living as pleasant as possible. A sustainable city will improve the quality of life in the city, including ecological, cultural, political, institutional, social and economic components, without leaving a burden on future generations.

Waste management

In 1999 the European Parliament brought in new legislation that waste was to be treated so that the amount of waste sent to landfill would be reduced. Landfill sites across Europe were required to change the way that they operated, complying with national and EU policy.

The waste hierarchy

The waste hierarchy (Figure 12) is seen as being the cornerstone of EU waste policy and legislation and is a core principle of the strategy. Its main purpose is to minimise

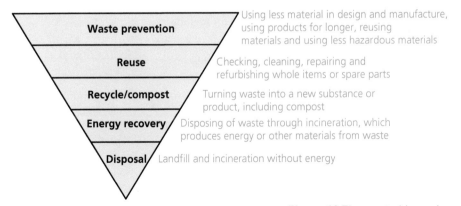

Figure 12 The waste hierarchy

> The **carbon footprint** is the total set of greenhouse gas emissions caused by an individual, event, organisation or product expressed in its carbon dioxide equivalent.

Knowledge check 9

What is the difference between a carbon footprint and an urban ecological footprint?

Exam tip

Be aware of the different ways that sustainable development in urban areas depends on the management of waste, energy and water.

the environmental effects that waste brings and to increase the resource efficiency in waste management and policy.

Energy consumption

There has been an increase in the number of technological devices that are powered by electricity. The demand for energy is increasing right across the world, with consumption expected to rise by 66% between 2008 and 2035. The demand for energy consumption will put pressure on the amount of living space that people have access to and this will also affect food production.

Many cities have started to realise that their capacity to create energy is at maximum levels and have therefore taken steps to try to reduce the demand for energy resources within the city.

The City Council of Edinburgh has taken big steps to reduce its energy use. It is keen to reduce carbon emissions and aims to achieve a 20% reduction in energy use by 2020 by using less energy to:

- light and heat buildings, tenement stairs and streets
- fuel vehicles
- run electrical equipment

Water supply

As cities are continuing to expand on an annual basis, the amount of water required to supply them is increasing rapidly. In 2000, there were 300 million Africans who were living in a water-scarce environment. By 2025 this is expected to double to 600 million people.

In Mexico City, over 21 million people live in an area around $1485\,km^2$. Much of the water supply has to be brought more than 100 km, 2,400 m above sea level. Supplying water to the city has become expensive, inefficient and energy intensive. Mexico City receives more rain than London but experiences huge shortages of supply. At every stage in the collection of water in reservoirs, its supply through pipes and through water treatment works, there is a series of conflicts with local people who feel that they do not benefit from the water strategy. Many locals who work to help supply water to the city cannot get a reliable, clean supply of water for themselves. They note that 'pipe pressure matches income levels' — people in many of the wealthier parts of the city can enjoy as much water as they want (and there is enough in the Miguel Hidalgo and Cuajimalpa districts for the golf courses). However, in the poorer regions (like Iztapalapa) taps can often run dry.

> **Knowledge check 10**
>
> Why can water supply be such a sensitive issue in some cities?

The SWITCH project took place from 2006 to 2011 and investigated 'Sustainable Water Management in the City of the Future'. This was an action research programme funded by the EU that aimed at catalysing change towards a more sustainable urban water management programme (Figure 13).

As a result of the five-year programme and investigation, a number of more sustainable approaches across the different aspects of water management were suggested (Figure 14).

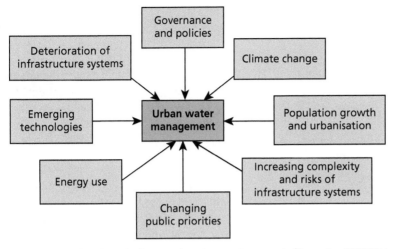

Figure 13 Issues and further challenges in urban water supply (from the SWITCH report)

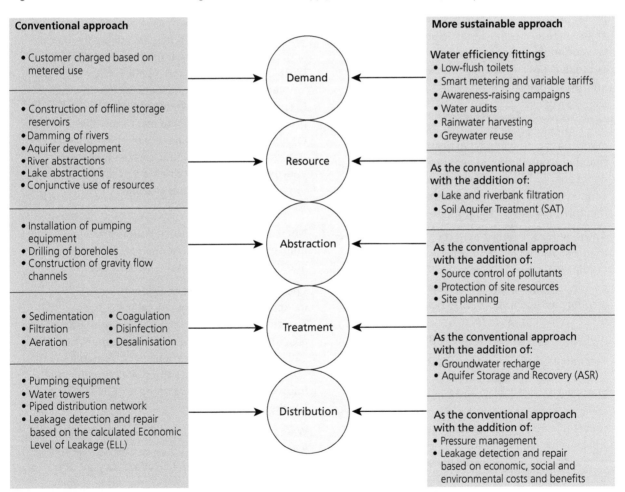

Figure 14 Water supply management components (from the SWITCH report)

Sustainability in a city

In 2017, the Belfast Agenda became the latest in a series of government-backed planning documents that would take a look at how public services work together within the community to deliver improvements for local people. The Belfast Agenda was created through a partnership of key city partners, residents and community organisations and has developed a vision for the city of Belfast for 2035.

Their intention was to establish Belfast on a sustainable development framework, 'making the necessary decisions to realise our vision of economic growth, maximising wellbeing and protecting the environment, without affecting the ability of future generations to do the same'. There is a commitment to make sure that economic success is not constrained by issues related to energy, transport and city water infrastructure.

Waste management

The Belfast Agenda notes that, 'Successful cities manage the impact of growth and ensure that it does not limit the quality of life of future generations. This includes reducing our consumption of non-renewable resources and minimising and managing waste effectively.'

Any plan for the future of Belfast needs to ensure that the city can meet all of its recycling and landfill obligations. Belfast City Council (BCC) has developed a comprehensive waste management strategy in recent years. BCC has joined with five other council areas to form arc21 to control and manage waste in a sustainable manner.

The adoption of the Revised Waste Framework Directive (WFD) in 2008 provided a foundation for sustainable waste management practice. One key element of this was the move towards a more recycling society that would:

- achieve a recycling rate of 50% (including preparation for reuse) of household waste by 2020
- achieve a recovery rate of 70% (including preparation for reuse, recycling and other material recovery) for all non-hazardous waste by 2020.

BCC has invested money into supplying residents with a range of kerbside bins that are collected regularly and into which they can separate their waste. In 2013–14 BCC collected 146,000 tonnes of waste and was able to recycle or compost 36% of this.

Energy consumption

The Belfast Agenda notes a commitment to work with partners in order to manage energy use across the city more effectively, to invest in renewable sources of energy and to tackle fuel poverty. The contribution of renewable energy in NI reached 25% in 2016. The major source of renewable energy in NI is wind power (91%), with 4% from landfill gas, 1% from biomass, 3% from biogas and 1% from other sources. The Local Development Plan 2020–2035 (2016) for Belfast sets out a policy that will help to 'implement renewable targets for the city that aim to reduce carbon emissions, while encouraging the promotion of its citizens to use renewable energy taking account of the impacts of climate change'. The report recognises that current patterns of energy use are unsustainable and have a huge impact on environmental change.

Water supply

Northern Ireland Water (NI Water) provides water and sewerage services to the 655,000 households and 85,000 non-households across Belfast.

The sewerage network in Belfast was built in Victorian times and was suffering due to sustained underinvestment. NI Water undertook a £160-million storm water management project called the Belfast Sewers project from 2006 to 2010. This would reduce the risk of flooding and improve water quality in the River Lagan and Blackstaff River.

Water supply in Belfast has long been a contentious issue. Decisions were taken to transport water from the Mournes area to Belfast through the Belfast Water Act of 1893. Water intakes were constructed on the Kilkeel and Annalong rivers and pipes and tunnels were constructed to move the water up to Belfast. A second stage

→

of development started in 1923: Silent Valley reservoir was constructed to quench the demand for increased supply. More recently, £90 million improvements were put in place in 2014 to lay new water mains throughout NI. Over 65 km of pipes were laid throughout Belfast so that the quality, reliability and security of the local water supply was maintained.

Summary

- Sustainability is the key ingredient of the UN SDGs for 2030, especially in relation to Goal 11.
- There are a range of social and environmental considerations in relation to sustainable development.
- Urban ecological footprints measure the amount of land required to produce the resources needed by one person, to support their lifestyle.
- Carbon footprints are the total amount of greenhouse gas emissions measured by the amount of carbon dioxide and methane gas emissions.
- The sustainability of an urban area depends on the management of waste, energy and water.

Urban planning and design in relation to sustainability

Urban planning considers all aspects of land use including transport, communications, housing, commercial and industrial buildings and distribution networks.

Urban design, planning and management in eco-towns or cities

In 2007 the British Labour government announced plans to develop 15 new eco-towns across England. In 2009 a new planning policy statement was made in the UK that would help to roll out the standards that eco-towns would have to meet.

The main aspects of this policy were set out in the Eco-towns: supplement to Planning Policy Statement 1 (2009). The government's main objectives for planning as set out in PPS1 include: To promote sustainable development by 'ensuring that eco-towns achieve sustainability standards significantly above equivalent levels of development in existing towns and cities by setting out a range of challenging and stretching minimum standards'.

Eco-cities are seen as being cities that are built from the principles of living within environmental means. They will usually adopt a carbon-neutral policy, use renewable energy resources and attempt to manage city life with the natural environment. The ideal 'eco-city' should have the following requirements:

- Operates within a self-contained economy. All resources required will be found locally.
- Will be completely carbon-neutral and use renewable energy resources.

Urban planning is concerned with the development and land use, protection, environment, public welfare and design of urban spaces.

Exam tip

Make sure that you have a good depth of understanding of how eco-towns and cities are managed differently from 'regular' housing developments.

- Will have a well-planned layout and public transportation system that prioritises walking, cycling and then public transportation.
- Will conserve resources — maximising the efficiency of water and energy resources, developing a waste management system with an emphasis on recycling.
- Will restore environmentally damaged areas.
- Will ensure that there is access to quality, affordable housing for all aspects of society, and will aim to improve job opportunities for disadvantaged groups.
- Will support local produce and agriculture.
- Will promote a simplicity of life — with an awareness of environmental issues as the central focus and concern in decision-making.

Knowledge check 11

What are the key aspects that allow for the development of an eco-town?

Urban design

Urban design is more than architecture, which focuses on the design of an individual building or structure. It looks at the larger scale and considers groups of buildings, streets, public spaces, neighbourhoods, districts and cities. The purpose of urban design is that urban areas will be functional, attractive and sustainable places.

Urban design is the process of designing and shaping cities, towns and villages.

PPS1 notes that when trying to identify a suitable location for eco-towns the following need to be considered:

a   The area for development needs to be able to provide for at least 5,000 homes.

b   The proposed eco-town needs to be close to a higher order centre, where there is a clear capacity for public transport links.

c   The eco-town needs to be close to existing and planned employment opportunities.

d   The eco-town might be able to play a role in helping with other planning, development or regeneration objectives.

Planning

The rules and regulations surrounding the standards expected in an eco-town can be very detailed. Some of the key features that planners need to consider when planning these developments include:

- Zero carbon
- Climate change adaption
- Homes
- Employment
- Transport
- Healthy lifestyles
- Local services
- Green infrastructure
- Landscape and historic environment
- Biodiversity
- Water
- Flood risk management
- Waste

Exam tip

Look into each of these aspects of planning development in more detail. How are they different from the management and design of traditional housing sites?

Management

Every eco-town is expected to have an overall master plan that will demonstrate how all of the planning standards listed above are to be adopted and achieved in the long term. Eco-town developments are monitored carefully by regional planning bodies and the local planning authorities.

North West Bicester is the UK's first eco-town. It was given the green light in 2009 and received planning approval in July 2012. The NW Bicester website describes the venture as 'a vibrant flagship project bringing investment, homes and jobs to the town.

It will create resilient, safe and strong communities and provide desirable homes that inspire and empower people to achieve a better lifestyle.'

The first phase of the plan of North West Bicester, Elmsbrook, has 393 zero-carbon homes, a primary school, a local shop, an eco-pub and a community centre. There will be 119 affordable homes available for rent or shared ownership. The plan also includes incorporating 40% of land for wildlife and biodiversity and 17,500 m² of solar panels, creating the UK's largest domestic solar energy array.

Urban design, planning and management sustainability

The world is urbanising at a very fast rate. By 2050, it is expected that two-thirds of the world population will live in urban areas. The number of urban residents is growing by around 73 million every year. The 2030 Agenda for Sustainable Development turns its attention to this challenge as SDG 11 aims to 'make cities and human settlements inclusive, safe, resilient and sustainable'.

The UN recognises that over 3.5 billion people already live in cities and that 95% of the urban expansion of the future is going to take place in LEDCs. The number of people who live in slums today is estimated at 828 million and this figure is going to increase rapidly over the next 20 years. Cities also occupy 3% of the Earth's land but make up 75% of the Earth's energy consumption and 75% of its carbon emissions. Urbanisation continues to put pressure on resources and there are concerns about the impact that further expansion will have on water supply, sewerage, the living environment and public health.

Residential space

The UK is currently undergoing a housing crisis. Newspapers have been reporting a catastrophic slump in home ownership and indicate that prices have jumped 151% since 1996, even though real earnings have only risen by about 30%. An estimated 1.2 million people are on housing waiting lists in England alone and another 6 million face tenure insecurity with no prospect of being able to purchase their own home. In 2015 around 140,000 new homes were built, even though the required supply per year is closer to 300,000 units.

Successive governments have been encouraged to invest in the housing market and to help first-time buyers into the property market. But funding is not the only issue; any increase in house building must also consider planning permission, green belts, new towns, developers' land banks, inefficient building methods and minimum space standards.

Housing design

In the nineteenth century, factories were built close to the city centre (or CBD) as this is where the best transport links could be found. Rapid urbanisation brought workers who were housed in cheap and quickly built terraced houses. Their managers might have found accommodation in slightly larger, detached houses along the main arterial routes.

The interwar period (1919–1939) brought change to the urban landscape. Local authorities became involved in building homes for rent. In parallel, private house building grew rapidly as mortgages became more affordable. Cities expanded and grew beyond the inner city into the suburbs of the city. The design of the houses changed too,

> **Exam tip**
>
> As cities expand, there will be more pressure for things to be done differently so that sustainable development is kept to the fore.

from two-up two-down terraces with outside toilets to more spacious semi and detached houses that had indoor plumbing, bathrooms and even space for garages and gardens.

Following the Second World War (1945–1959) a new period of building responded to the destruction caused by the war. Prefabrication was used as a way of supplying new homes quickly. Cities continued to sprawl into the countryside and the footprint for the houses continued to expand. In more recent times, sub-urbanisation and counter-urbanisation have encouraged house building in the suburban areas and beyond. Houses have continued to increase in size and modern living requires that people have space for cars, energy for heating and good electricity supplies.

Defensible space

Oscar Newman had been developing ideas in relation to crime prevention and neighbourhood safety in the city. He argued that the design of any place should 'allow inhabitants to become key agents in ensuring their security'. His theory argued that an area would only become safer when the residents felt a real sense of ownership and had developed a shared responsibility for the community. Newman wrote that 'the criminal is isolated because his turf is removed'. The idea is that people will feel comfortable challenging any anti-social behaviour or crime in their immediate areas. There should be a sense of a watchful community that will cause criminals to think twice before operating in an area. This can be further extended through the adoption of neighbourhood watch groups.

However, Newman noted that any family's claim to a territory would diminish as the number of other families in the area increased. In other words, they would have less control over the area. The more people there are sharing a communal area, the more difficult it is for people to feel a connection or responsibility for it.

Greenfield and brownfield development

Green belt policies are often put in place to actively prevent any urban sprawl into greenfield areas. Planners use the concept of a green belt as an invisible line to prevent urban development into an area, seeing the green belt as the 'lungs' of an urban landscape.

In November 2016 the Royal Town Planning Institute (RTPI) suggested that England's greenfield sites, including green belts, should be considered along with brownfield land as locations for new housing. In their policy statement called 'Where should we build new homes?' they argued that the government needed a fresh approach to housing policy.

People from the suburbs or from outside the city might move back into the heart of the city. This is called re-urbanisation. The stimulus for any movement usually comes as a result of government investment and intervention to improve the city. The development of the Titanic Quarter was an attempt to re-urbanise parts of Belfast that had been brownfield sites.

Retail parks

Over the last 30 years, the number of out-of-town retail parks has increased. Modern retailing has noticed the growth of large out-of-town superstores located close to residential areas where free parking allows people to park and shop in relative

Exam tip

Think about how the balance of housing types might be changing in the city and why.

Knowledge check 12

Why has housing design changed so much in NI over the last 120 years?

Greenfield development refers to an area of land surrounding a city or town that has not been developed or built up.

A **brownfield** site is a piece of land that has already been used and is now lying derelict.

Re-urbanisation is the movement of people back into an area that had previously been abandoned.

Knowledge check 13

Why would most governments prefer to develop brownfield rather than greenfield sites?

comfort. Other shops that cater for DIY, technology and furniture might also cluster together to form a retail park or mall experience. The more traditional central shopping areas or high streets in towns and cities have increasingly come under pressure. Corner shops have all but disappeared. There are a number of reasons for this change in shopping patterns:

1 The more affluent people in society have become suburbanised and now live on the outskirts of the city.

2 Technological changes mean that more people have access to personal cars and allow a flexibility of shopping.

3 Economic changes mean that people have more disposable income and can now shop on a more regular basis.

4 There is much congestion in city centres so people prefer to stay out of the city.

5 The out-of-town shopping centres are often more accessible with free parking and access to a wide range of shops, food and entertainment under one roof.

6 Social changes mean that as more women are working, shopping has become something for both sexes and has become an important part of leisure time and activities.

Table 9 Environmental and social consequences of the development of retail parks

Environmental positives	Environmental negatives
Land on the edge of cities/towns is cheaper which means that developers will have more money to spend, making buildings more sustainable.	If land is cheap, the environmental 'footprint' of these retail parks can be large.
Building these parks on the edge of cities can reduce congestion/traffic jams and pollution build-up in the city.	Greenfield sites are much easier to develop at the edge-of-city locations than brownfield sites in the inner city.
Many of the retail parks are developed to fit in with their surroundings and often make the area more attractive (with landscaping).	Most retail parks are planned with huge car parks. These encourage the use of personal motor vehicles which are much less environmentally sustainable. Public transport links are not as good to retail parks.
Social positives	**Social negatives**
Many large developments offer free parking for visitors.	Traffic and congestion will now surround these retail parks at peak times and on Saturday; it can create problems for local people.
The development of retail parks will improve the local transport infrastructure; new roads, bus links and maybe even rail links will be built.	The retail parks are set up to draw people from a wide sphere of influence. They are not aimed at a local audience and the opening times might not suit the local person.
Many of the retail parks include social areas — coffee shops and charity shops as well as more specialist types of shop. Modern shops also have play parks, cinemas and child care facilities.	Local people will also be the first to notice the negative impacts of retail parks on their town centres. Shops will face increased competition and may close as they struggle to compete with national retail giants.
A wide variety of jobs and retail opportunities are set up for local people and their children.	

Exam tip

Make sure you understand the positive and negative consequences of both the environmental and social factors involved in the development of retail parks. How do these compare with retail in city centre locations?

Leisure and sports facilities

Urban design and planning is required to ensure that the social facilities required for modern life are built into any development.

In 1902, Ebenezer Howard wrote an article called 'Garden cities of tomorrow'. Howard was a social reformer who tried to summarise the political, economic and social context that lay in his vision for the future of British settlement. His diagram 'The Three Magnets' (see Figure 15) looked at the differences between the magnets of town and country life and found a third magnet — the town/country or the garden city — as the best of both worlds.

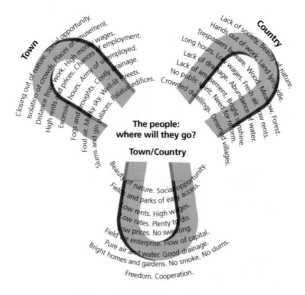

Figure 15 The Three Magnets

The World Health Organization (WHO) note the important part that urban green spaces play: 'Green spaces such as parks and sports fields as well as woods and natural meadows, wetlands or other ecosystems, represent a fundamental component of any urban ecosystem.'

Urban open space and urban parks are open areas that might be playing fields, recreation areas, parks, open countryside, piazzas, plazas and urban squares. In most places these are usually green spaces. These can be important because:

- **Recreation**: These spaces will allow people in cities to have daily access to nature and can provide opportunities for peace and relaxation in the busy-ness of city life. Time spent in an urban open space can offer a reprieve from the urban environment.
- **Ecological positives**: Nature can be conserved in the city. People have contact with nature in the urban surroundings which can promote biodiversity. Plants and animals can survive in the city that would not otherwise be able to survive. Trees also help to filter out soot and other pollutants from the atmosphere. They also help to produce oxygen.
- **Aesthetics**: People like to have good views and nature is a welcome relief from the urban landscape. This can have a positive psychological influence on how people view their life in the city.

> **Exam tip**
>
> Make sure you have a good understanding of why it is important for leisure facilities and open spaces to be part of any urban plan.

- **Public health**: Physical activity, such as walking, jogging and following trails through the green space help people to live healthier lives.
- **Play space**: Spaces are provided that allow children to play, especially in areas where there are no gardens.
- **Mental health**: WHO recognises the contribution that green spaces can play in relation to mental health: 'having access to green spaces can reduce health inequalities, improve well-being, and aid in treatment of mental illness. Some analysis suggests that physical activity in a natural environment can help remedy mild depression and reduce physiological stress indicators.'

Leisure and sports facilities are also an important part of any urban planning. It has been widely recognised that any new housing development will need to include some level of community or indoor leisure facilities to serve the local residents. These facilities can be commercial (private) or council (public) facilities such as health and fitness gyms, sports halls and swimming pools.

Case study

Urban planning and design in a city

The people of Freiburg, Germany consider their city to be the birthplace of the environmental movement. Over the last 30 years, Freiburg has taken measures to ensure that urban planning has allowed the city to be able to enhance the lives of the 225,000 people who live there. The city is a 'green city' which means that environmental policy underpins its planning policy at all levels.

Residential space

New housing developments in Freiburg have created safe, inclusive and sustainable communities. One place where over 6,000 people are now living is inner-city Vauban. The area was a former French military base and has been under development since 1999. In Freiburg, passive and energy-saving homes have become the norm.

The houses have been planned in such a way as to encourage people to prefer to travel on foot or by bicycle. The development is connected to a tramway. In 2009, around 70% of households recorded that they had chosen to live without a car. Houses are designed to be crescents or cul-de-sacs where a series of pedestrian and bike paths have been integrated to move people from one place to another. The paths go through open spaces.

A variety of sustainable techniques are built into the building design. Many of the houses are built to be cost-efficient and optimise the energy produced. Houses are equipped with vacuum toilets that dispose of waste products into a biogas reactor. The gas is then used for cooking. Waste water is treated in a sewage-treatment plant. Solar panels help to produce further electricity in a sustainable manner. Community-based infrastructure includes schools, nursery schools, youth facilities, civic meeting places and a market place. Vegetation-covered 'green' roofs will store rain water which can also be collected and reused in the district.

Greenfield and brownfield development

The urban planners in Freiburg recognise that open land is a resource that is quickly vanishing. In Germany the amount of land use per head of the population keeps going up. This means that inner-city development has had the highest priority in Freiburg. They want to ensure that they do not develop every brownfield space in the city but concentrate on some areas. The development in Vauban is a good example of how a brownfield site has been used for residential development. The area is dominated by four-storey buildings that offer a sense of community and defensible space.

➡

The Rieselfeld development to the west of Freiburg is based on an area of land that used to be a sewage works. The city has always had an inwards rather than outwards policy in relation to the use of brownfield sites instead of using green space.

Retail parks

In Freiburg a number of the shops moved to more out-of-town locations in the 1980s. Three major shopping malls/retail parks were developed: Breisgau Center, ZO – Zentrum Obserwierhre and West-Arkaden. These are easily accessible to pedestriains, cycle paths and tram routes and shoppers are encouraged to use public transport instead of taking cars to the shopping centres. This helps the city to attempt to make the impact of the shopping centres as sustainable as possible.

Leisure and sports facilities

Urban planning policy in Freiburg notes that urban planning not only impacts built-up areas but also the open landscape. Planners note that open landscapes have a variety of functions: 'living space for plants and animals, a protective environment for the resources of earth, air and water, and a recreational space for people. Green spaces between the rows of houses allow for good climatic conditions and provide play areas for children. The Freiburg planners are committed to the "maintenance of parks and public green spaces for their environmental and climatic value".'

The city has become famous for its green spaces: parks, landscape conservation reserves, nature reserves, garden plots, playgrounds for children and cemeteries. The green areas help to add value to the areas; they improve the microclimate and habitats for plants and animals. Of the 152 open spaces in Freiburg there are 64 playgrounds, 28 soccer fields and 19 ball game courts. Parks such as Seepark and Dietenback Park offer space close to densely populated areas.

Summary

- Eco-towns and cities require careful urban design and planning measures to ensure that their levels of sustainability are managed carefully.
- The way that urban areas are designed, planned and managed are linked to sustainable development through:
  - the development of residential space and how houses are designed and seen as providing 'defensible space'
  - the desire in most cities for fewer greenfield sites to be developed but for further development of brownfield sites
  - a broad range of environmental and social positives and negatives will result from any development of out-of-town retail parks compared with any plans to maintain retail in the city centre
  - ensuring that leisure and sport facilities, open spaces and urban parks are part of any plan for urban design is important for the health of the city and the people who live in it

Traffic and transport

The movement of goods, materials and people takes a lot of energy and has a big impact on the environment, using up the Earth's valuable resources in the process.

Impact of sustainability on transport

Sustainable transport includes the social, environmental (and climatic) impacts that moving materials and global supply cause. Sustainable transport should then:

- allow for the basic needs of individuals, companies and society to be met safely
- be affordable, operating fairly and effectively across a variety of transport modes
- limit emissions and waste as much as possible and use renewable resources where possible

Exam tip

Make sure you have a good understanding of what makes up sustainable transport and how countries are moving away from carbon-based transport methods to more renewable methods.

The five elements of sustainable transport can be viewed as:

1 **Fuel economy**: Each type of transportation will have a different impact on the environment and its 'carbon intensity' (Figure 16). By improving fuel economy we can get the same mileage while generating fewer emissions.

2 **Occupancy**: One of the easiest ways to lower the carbon intensity of a passenger kilometre is to make sure that vehicles are fully occupied.

3 **Electrification**: There has been a lack of breakthrough in biofuel technology. This means that many consider electrification as the most important path to achieving a low carbon transport cost.

4 **Pedal power**: Bicycles have a very low carbon footprint as they do not use traditional energy sources or contribute to greenhouse emissions. However, this does limit transport to a local scale.

5 **Urbanisation**: People who live in cities have lower transport emissions. They drive less, use more public transport, cycle and walk more. However, in some ways this is offset by the amount of energy required to transport food and materials into the city.

Knowledge check 14

What can help to make transport more sustainable?

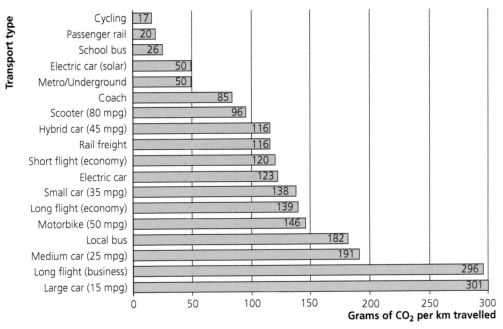

Figure 16 The carbon intensity of transport

Transport by sea

Much of the modern globalised trade network was initially based on sea travel. Ports in the UK and Europe have developed links across the Atlantic Ocean and further afield for many years. Rapid development and urbanisation was fuelled through the importation of cheap raw materials from LEDC countries that would be manufactured into goods to be sold internationally. In recent years, the amount of goods exported from British ports has declined but the amount of manufactured goods from other countries has boomed (Table 10). Ships have increased in size so that special bulk carriers and massive container ships will travel the globe, moving materials from one country to the next. Ships stay in port for as short a time as possible to maximise the number of miles they travel per year.

Table 10 UK major port international imports and exports (2000–2015)

Year	Imports (million tonnes)	Exports (million tonnes)
2000	200	190
2005	300	170
2010	260	160
2015	285	152

How sustainable is sea transport?

Sea transport is a slow but economical way of transporting things from one place to another. The amount of expense increases with distance, though the cost of maintaining and updating ships with the most up-to-date technology or passenger facilities can be expensive. Port costs and taxes can be expensive. Freight facilities can offer good value for money and can be the cheapest method for moving goods from one place to another. However, routes and schedules can be highly restrictive. There are only a few ports that are able to cope with the bulk delivery of goods into the UK.

Environmentally, sea transport does mean that ships can be used for many years. However, the ships can also transport disease, pollute water and leak oil. When they crash, there can be huge environmental consequences resulting in expensive clean-up operations. They also require large amounts of land for storage and links with additional transport methods to transfer the goods or people on.

Transport by air

Air transport has become one of the most frequent ways to travel internationally. It more than doubled from the 1980s to the year 2000. Although air transport has been around since the 1900s, the commercial aviation boom really only took off following the Second World War. In recent years the number of passengers travelling through UK airports has increased hugely. The number of passengers who travel through London Heathrow airport alone each year exceeds the total population of the UK.

Air travel is also used to transport freight that needs to arrive quickly at its destination. The amount of freight does not compare with the amount of freight put through the sea ports but it still adds a valuable contribution to the economy of the country.

How sustainable is air transport?

Air travel can be one of the least sustainable methods of travel. The only mode of transport that can create more CO_2 per kilometre travelled is the large motor car. A report in 2014 from the US Department of Transportation notes that global carbon dioxide emissions due to commercial aviation will reach 1.5 billion tonnes per year by 2025. There is much research at present to overcome the many technology challenges that would allow air travel to be more sustainable. Motors, fuel cells and batteries are all being tested in the hope that the older jet fuel guzzling engines can be replaced and use renewable technologies such as liquid hydrogen.

Air travel is fast, especially over longer distances, but sometimes the travel times over short distances can be inflated due to delays. Air travel is expensive and really is only accessible by those with money. Airports can be controversial; plans to add another runway to London Heathrow have met fierce resistance from the local people. Landing fees and airport taxes can push prices up. The actual planes are also expensive to buy or lease and they need to be in constant circulation in order to make money.

Environmentally, there can be high noise levels along the aircraft flight paths and huge amounts of land are required to supply the space required for modern airport facilities. However, the single biggest factor is the air pollutants that are given off by the aircraft. In the EU between 1990 and 2005, the amount of greenhouse gas emissions by aviation increased by 87%. Efforts in the aviation industry have been made to reduce CO_2 emissions and make air travel more sustainable. The Airbus 380 is a new plane that was designed specifically to be a 'green giant'. However, even with this design the fuel performance of the A380 has been tested and shown to be broadly similar to a Boeing 747 or 15% worse than a Boeing 777.

Transport by land

The UK government recognises that its role is to support the transport network that helps the UK's businesses and gets people and goods travelling around the country. Transport by land is the most local and the most direct and flexible type of transport. It is also the most individual as most adults in the UK have access to their own vehicle. Some figures from the UK Office for National Statistics indicate that:

- The total distance travelled by people in Great Britain (GB) increased by 275% between 1952 and 2007 to 817 billion passenger kilometres per year.
- The proportion of households in GB without access to a car decreased from 86% in 1951 to 22% in 2008. In addition, the proportion of households with access to two cars increased from 1% to 27%.
- In 2007 82% of GB passenger travel involved cars/vans. Bus/coach and rail travel were around 7% each, with other modes (walking, cycling and domestic flights) coming in at 4%.

How sustainable is land transport?

The UK government has a number of key sustainability targets that it needs to reach, including the Climate Change Act 2008 to reduce greenhouse emissions by 80% in 2050. Around 94% of all greenhouse gas emissions in the UK come from road transport.

There has been much research and development recently into more sustainable forms of transport. Green vehicles are designed to have a reduced environmental impact compared to standard vehicles. Electric vehicle technology has developed in recent years. Some cars such as the Nissan Leaf can be charged directly through electric cables. Hybrid vehicles will use an internal combustion engine and an electric engine to achieve better fuel efficiency. Natural gas and biofuels are also sometimes used in vehicles but these have wider consequences.

Canal and river transport

Rivers and specially built canal waterways were one of the first ways of transporting goods en masse within the UK. This can be a slow method of transport but it can also be cheap as very little fuel is used. The use of canals hit a peak during the Industrial Revolution but many canals have fallen into disrepair and would require much maintenance for their continued use. There are relatively few environmental issues in relation to canals, apart from perhaps the displacement of habitats that might be caused during their building. Today, canals are largely used for leisure purposes.

Exam tip

Air and sea transport are relatively straightforward and while road transport (passenger car and freight) is the most dominant, you should also consider the role that canals, rivers and rail might play.

Rail transport

From the mid-nineteenth century railways began to replace canals. By 1830, the first intercity railway route (between Liverpool and Manchester) was opened. Further railway lines developed quickly throughout the British Isles and passenger numbers grew quickly.

Since 1995 there has been a massive resurgence in the importance and use of the railway. The UK currently has the seventeenth largest railway network in the world and in 2016 there were 1,718 million passengers. This makes it the fifth most used railway network in the world. Rail travel can be quite fast and can be relatively cheap to operate though the number of people required to work on the railways can be high and this can increase costs. Development costs (such as for the proposed HS2 line in England) can be very high but can be cost-effective in the long term. Trains can be used for the transfer of heavy, bulky goods, mail and passengers.

Environmentally, rail transport does mean that noise and visual pollution are low and only really occur in specific places. The use of fossil fuels for energy does cause carbon emissions but these are much less than if the same number of people were to travel by road. New electric trains can be much more environmentally friendly. In addition, the engines and rolling stock can last for a long time which makes them more sustainable than other types of transport.

Road transport

Since the late 1960s in the UK, the motor car has become the most important mode of transport. Over 90% of all journeys are taken in personal vehicles. Road transport can include vans and taxis, buses and motorcycles, but the vast majority of people rely on the car. In addition, 65% of domestic freight is carried on the roads.

Environmentally, road transport is considered one of the least sustainable. Cars are rarely full of passengers. People like to have their flexibility and this means that the number of cars on the roads of the UK continues to increase each year. This causes huge impacts on carbon emissions and on global warming. Roads also take up valuable green space and can kill animals.

Evaluation of urban traffic management strategies

Governments around the world have found it necessary to come up with a response to the ever-increasing issue of traffic congestion in urban areas. The establishment of effective **traffic management strategies** has become a central issue in modern urban planning.

Traffic management strategies attempt to reduce the use of private cars and encourage people to look for alternative solutions to getting around.

Public transport strategies

The most successful way of managing urban traffic is to try to encourage people to use public transport methods more. Buses and trains are more sustainable as the engines can move larger numbers of people which is not only better for the environment but will also take an increased number of personal cars off the roads. Governments need to invest money into the infrastructure and ensure that there is a reliable, regular and cost-effective service that meets the needs of the people.

Many large cities have developed underground mass transit systems that can move huge numbers of people on a daily basis. The London Underground carries around 1.3 billion passengers each year, or 4.8 million passengers each day. It has 270 stations and over 400 km of track. As a result, car ownership in London is relatively low; people are more likely to be regular users of the Underground system.

Integrated transport networks

As public transport systems become more widespread, they can combine with other modes of transport to provide a more holistic service. Bike racks and storage facilities allow people to cycle part of their journey and then to change transport mode. In London, large bus stations (such as Victoria) and the railway and underground stations are all located together so that travellers can change quickly.

Restrictions on car usage

Some cities such as Freiburg have put severe restrictions on the use of any form of personal transport. Cars are not permitted into the city and there are hefty charges for people to hire a garage at the edge of the city. As a result, there has been a rise in the number of car rental and shared ownership schemes. People might only require a car for three days per month so they will book a car ahead of time and share the ownership costs. Congestion charges are a more common feature of cities. In London, fees were introduced from 2003 that would charge £11.50 each day for each non-exempt vehicle that travelled into the central zone. There are also plans to charge an additional £10 to older polluting diesel vehicles in the future. In 2013, Transport for London reported that the congestion charge scheme has helped to reduce traffic volumes in London by 10%.

Car parking

Over the years there have been a number of car parking policies in Belfast that attempted to take pressure off the congestion in the CBD and inner city. Initially, cities might allow car parks in the centre to encourage people into the city, but these were found to create more chaos. Multi-storey or underground car parks were then built a little further out of the city centre on or close to roads that led out of the city. Park and ride schemes originally brought people from the edge of the city into the centre but these have been widened so that people can park up to 10 miles out of the city and get a bus directly into it. Parking capacity has also increased at railway stations outside the city to cope with the massive numbers of people now wanting to use public transport to beat congestion. Parking in the city centre is now very expensive.

Pedestrian and cycling policies

Many cities have adopted an approach where drivers are encouraged not to use their vehicles but to walk or cycle to work. Obviously, this works better with people who live and work in the same area. However, cycle lanes can also be used to speed up travel for people who live further afield. In Belfast, new cycle lanes and an investment in the rental facility Belfast Bikes has shown an increase in the number of people who cycle through the city. There are around 20,000 rentals of these new bikes each month. In addition, new urban greenways such as the Comber Greenway connect a 7-mile stretch of cycle and pathway from Belfast to Comber.

Integrated transport networks are created when different types of transport are coordinated to make the transfer from one to another much easier, for example, train times and bus times might match up as they arrive into the same station.

Pedestrian policies can also be used to stop traffic from coming into city centres and other urban plazas. However, access can be limited for businesses to times at the start or the end of the day. It might be difficult to get into the areas as transport links and car parks are located outside the area. People with disabilities can find it more difficult to get to the places they want to. The reality is that pedestrianisation rarely solves the problem of congestion but just manages to shift the problem further out of the city.

Case study

An urban traffic management strategy

In 1969 the city of Freiburg in Germany developed its first General Urban Transport Policy where traffic and transport around the city would be managed in a way that would not affect urban development, nature or the environment (Figure 17).

Public transport

Freiburg operates a successful public transport system operated by VAG Freiburg. This consists of a network of electric trams, first started in 1983, that works right across the city. The first transferable monthly environmental ticket was introduced in 1984. The trams are electric and are now powered using sustainable sources (mostly solar and wind power). This means that the carbon emissions in the city are much lower than they were previously. Between 1982 and 1999 the use of public transport rose by 18% (Table 11). The number of people carried by VAG Freiburg increased from 27.3 million annual journeys in 1980 to 72.8 million in 2009.

However, the amount of transport links is still limited. It can take some time, and a number of connections, to travel from one side of the city to another. The operating times of the transport system can also be inconvenient for those who work shifts

Table 11 Freiburg's transport in 1999

Pedestrian	23%
Cycling	27%
Public transport	18%
Car sharing	6%
Other (mostly private car etc.)	26%

or unsociable hours. Often the public transport is overcrowded which can be difficult for travellers.

Integrated transport networks

Freiburg has adopted an integrated traffic management system (including pedestrian, bicycle and public transport) since the 1960s. Around 68% of trips in the city are now made using the trams. The public transport network is becoming increasingly integrated as cycle racks have been built at railway and tram stations to increase the seamless transfer of people from one transport mode to another. Some people choose to join car sharing clubs where they can book access to a car when they need to make big purchases or to travel to areas outside of the city. The city boasts that 'with new lines, scheduling and passenger comfort, the old tram systems

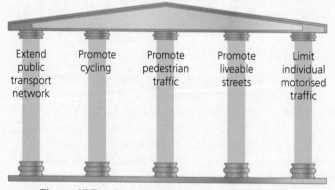

Extend public transport network Promote cycling Promote pedestrian traffic Promote liveable streets Limit individual motorised traffic

Figure 17 The five pillars of Freiburg traffic policy

have evolved into a modern city railway system that today serves almost all of the major districts of Freiburg'. Over 70% of residents in the city live near a tram stop.

Pedestrian and cycling policies

Freiburg has a large pedestrian zone in the centre of the city where no cars are allowed. The only way that people can get into the heart of the city is to use bicycles or public transport. There is much less air pollution around the city which means that people can enjoy being outdoors much more. There has been growth in the number of cafés and restaurants where people can sit outside, which helps the city to look and feel better. Shops and services are built together in different neighbourhoods so that shopping and medical services are located close to each other and can be visited on foot.

Cycling is one of the most used transport methods in the city. Between 1982 and 1999 the percentage of bikes used as the main mode of transport in the city rose from 15% to 27%. There are over 400 miles of cycle track across the city and bikes have priority on roads and at traffic lights. There is safe bike storage at multiple locations across the city (with over 6,000 bike parking slots). The first bike map was created in Freiburg in 1970. Today, bicycles are seen by many as the best way to travel and are supported by various buy-a-bike-for-work schemes and a network of bike parking garages and repair shops.

Restrictions on car usage and car parking

The use of the motor car is in decline within Freiburg. There are many areas of the city that are totally car-free. Areas of free parking have been removed and replaced with extremely high car parking charges. People who live in the Vauban area have to agree never to own a car. Those who do own cars find it difficult to park them in the city and have to pay £17,500 a year to the local council to park in a solar garage (multi-storey car park) at the edge of the city. The proportion of car journeys fell from 38% in 1982 to 26% in 2009.

People are discouraged from buying a car. This can mean that they have less flexibility, especially when they want to take longer trips out of the city. There are fewer carbon emissions; there has been a 35% reduction over the last 10 years.

Many of the shops and offices in the city still require vans and lorries to deliver materials. This is meant to happen in the early morning and late evenings but this can put more pressure onto the city centre workforce. Some jobs still require cars and the restrictions on road transport and traffic calming measures can cause problems for people who need to get round the city. In fact, the reality is that car ownership is still quite high around the city. Local people are not always in favour of restrictions on their choices and some feel that harsh restrictions put some people off coming to live and work in the city.

Exam tip

There are two potential exam questions you could use a case study of Freiburg for: the urban planning aspect and the traffic management strategy. Make sure you understand the subtle differences so that you do not get mixed up when you answer questions on this.

Summary

- Sustainable transport includes the social and environmental impacts that are caused through the movement of materials and people.
- Each mode of transport (sea, air and land) will have different advantages and disadvantages in terms of sustainability.

- An evaluation of urban traffic management strategies needs to look at a range of policies and developments: public transport strategies, integrated transport networks, restrictions on car usage, car parking, pedestrian and cycling policies.

Option C Ethnic diversity

The definition of ethnicity

Factors that define ethnicity

The concept of ethnicity is not one that can be easily defined.

Refer to information on page 8 about ethnicity.

Race

Refer to information on page 9 about race.

Nationality

For many people, the key aspect that unites is the shared experience of coming from the same place. Many people have a strong link with their country of origin and associate with the outward signs and symbols connected with this place, for example, national flags, anthems, customs and traditions. Nationality can be both unifying and divisive; when people feel part of the 'group' they have a sense of belonging, but when they do not feel part of the 'group' this can lead to a feeling of discrimination and can lead to conflict.

Nationality is usually identified as being part of the population of a nation-state, but sometimes nationality might be expressed as a hope to create a nation-state, for example, many of the Kurdish people in the Middle East identify themselves as Kurds, even if they live in Iraq or Syria. Ethnic groups will usually proclaim their shared identity and national unity to the outside world, for example, in Canada many people use the symbol of the red maple leaf as something which bridges the gap between the English- and French-speaking ethnic groups. In NI the usually Catholic nationalist population might see their national identity tied more with the Republic of Ireland than with the UK, whereas the usually Protestant unionist community might see their national identity as British and tied with the UK.

Language

Differences in language become apparent as people try to communicate with each other. There are over 6,500 different languages around the world. English is a widely spoken language (Figure 18) and around 33% of the world can speak it. The first language that a person speaks or writes is often directly linked to where that person has grown up and spent the first 10 years of their life.

Sometimes language is used to intimidate people or to belittle them. In NI, language is used as one way to show an outward expression of ethnicity. The nationalist/ republican community identifies strongly with the Irish language. Irish-medium schools have been built and Irish street signs have been posted in areas where people who share this identity live. In NI, the most recent census was not only translated into Irish but also into 10 other languages; this shows how NI is increasingly becoming more ethnically mixed. Often debates can rage about which language should be identified as the national language; in Canada there are frequent discussions about whether English or French should be the more important.

Exam tip

Ethnicity is not a simple concept to define, so make sure you look at the four main factors below.

Knowledge check 15

What is an ethnic group?

Exam tip

There is a lot of overlap in this section with some of the information required for Option A Cultural geography. You should refer to each of the sections and take notes on each of these key concepts.

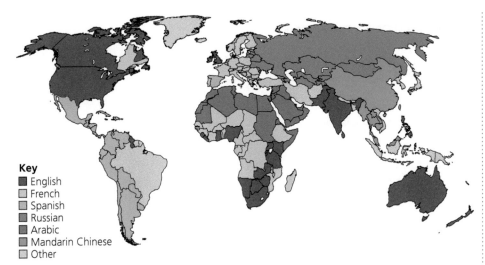

Key
- ■ English
- □ French
- □ Spanish
- ■ Russian
- ■ Arabic
- ■ Mandarin Chinese
- □ Other

Figure 18 Map of world language distribution

Religion

Refer to information on pages 9–10 about religion.

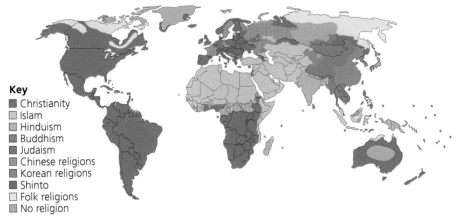

Key
- ■ Christianity
- □ Islam
- □ Hinduism
- ■ Buddhism
- ■ Judaism
- □ Chinese religions
- ■ Korean religions
- ■ Shinto
- □ Folk religions
- ■ No religion

Figure 19 Map of world religions distribution

Perceptions of ethnic and social identity

People can identify themselves easily and obviously with their race, nationality, language and religion. However, further factors can have a bearing on the particular ethnic or social group that someone might feel part of. This second list of factors does not impact people to the same extent and might influence them at particular times in their life. These factors might be less obvious and might be assumed of a person because of the perception of other people rather than what the people think themselves.

Role

The role that a person plays in society will have an impact on the perception they have of themselves or the perception that others have of them.

Refer to information on pages 10–11 about social class.

Exam tip

Any question that requires you to describe the factors that define ethnicity will also require knowledge of places around the world where ethnic factors were found.

Knowledge check 16

Why might perceptions of ethnic and social identity not be as important as the previous four factors?

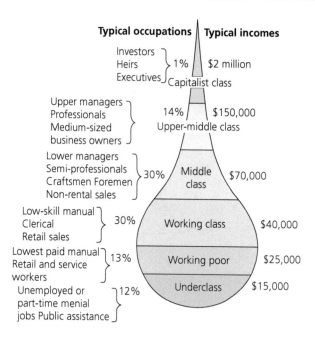

Figure 20 The social class system of the USA

In some parts of the world, the role that a person plays is dictated at birth. In India a caste system exists. It consists of two elements; varna is the class system which is broken down into four social classes:

- brahmins (priestly people)
- kshatriyas (rulers, administrators and warriors)
- vaishyas (artisans, merchants, tradesmen and farmers)
- shudras (labouring classes)

In addition, another group known as the dallits (untouchables) are usually removed from the varna system.

Residential concentration

International migration has long been involved in the expansion of urban areas around the world. As people move into the cities there is a tendency for those with similar ethnicities to group together in population clusters (enclaves). There is both a cultural and social reason for this, but usually it is due to a feeling of safety in numbers until the people feel that they have settled and integrated into the local society; they will feel safer being surrounded by people who have similar language, dress, food, nationality, race and religious beliefs. Services that are specific to the needs of this ethnic group will be established to serve the needs of the community and help them feel that the place is more like home.

For example, over the last 15 years there has been a significant increase in the number of people migrating from Nigeria to Dublin. Many of the immigrants initially move into an inner city area of Dublin, just south of the river Liffey, but as they become more settled over time they start to move towards the outskirts of the city.

Age

In some countries and cultures older people are given much respect. Many different tribal societies have village or local elders who rule and make decisions for the family group. Life experience produces wisdom. As life expectancies continue to rise, it is becoming more normal for people to survive into old age, which means that old people make up an ever-increasing sector of society. In many MEDCs this has started to put pressure on the resources of the country.

Some elderly people have noted discrimination due to their age; they find it difficult to get work in their advanced years.

Gender

Refer to information on pages 8–9 about gender.

Summary

- Ethnicity is not an easy concept to define but it can usually be defined in relation to race, nationality, language and religion.
- Ethnicity can also be defined by the perceptions of ethnic or social identity that an individual or group might hold. These can be related to role, residential concentration, age and gender.

Exam tip

Perceptions of ethnic and social identity also help to define ethnicity so make sure you are aware of how these can influence ethnic groups.

The processes that create and maintain ethnic diversity

Globalisation continues to be one of the most dominant forces in the twenty-first century. We live in a highly inter-connected world. People are on the move across our planet at rates that are faster than at any other time. Where obvious differences like race, nationality, language or religion might have once been a stumbling block, or cause for discrimination and abuse, these are now seen as strengths.

What is ethnic diversity?

Ethnic diversity is when there are multiple traditions/ethnic groups living in one particular area; different ethnic groups with different racial characteristics, nationalities, language or religious beliefs coexist. There will be ethnic mixing between the communities as they come into contact with each other. This enables the inclusion of individuals representing more than one national origin, race, religion or socio-economic group.

Over the last 20 years the ethnic diversity of NI has become increasingly complex (Table 12).

Knowledge check 17

What is ethnic diversity?

Table 12 Percentage of all usual residents in ethnic group

White	Chinese	Irish Traveller	Pakistani	Bangladeshi	Other Asian	Black Caribbean	Black African	Black other	Mixed	Other
98.21%	0.35%	0.07%	0.34%	0.06%	0.28%	0.02%	0.13%	0.05%	0.33%	0.13%

The 2011 NI census noted that:

■ 1.8% (32,400) of the population in NI belonged to one of the minority ethnic groups. The main groups within this were Chinese (6,300) and Indian (6,200).

■ 40% of the population claimed a British Only national identity, 25% claimed to be Irish only and 21% had Northern Irish Only as a national identity.

Processes that create ethnic diversity

The process of becoming an ethnically diverse society, with ethnic mixing and more than one obvious ethnic group, can be long and complicated. Each regional and local area will have been on its own particular journey with its own unique history of how it reached that point.

Colonisation

Colonisation is when one country takes political 'ownership' of another area or country. Often the process starts with the setting up of new settlements and the immigration of people to these areas, which will gradually develop and spread the influence and control of the new population.

Modern colonisation began from the fifteenth century as explorers were sent from the main western European countries to find, exploit and colonise new lands in the Americas, Africa, Asia and Oceania.

> **Exam tip**
>
> Make sure that you have a good understanding of what ethnic diversity is.

> Colonisation is the establishment and maintenance of rule, for an extended period of time, by a sovereign power over a subordinate and alien people that is separate from the ruling power.

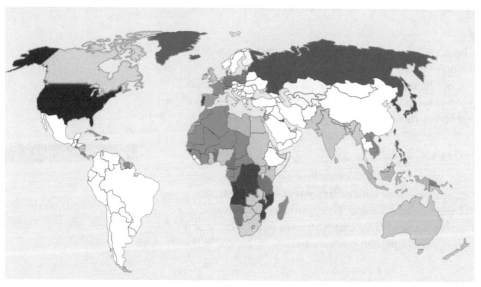

Key
■ Belgium
■ Denmark
■ France
■ Germany
▨ Great Britain
▧ Italy
■ Japan
▨ Netherlands
■ Portugal
■ Russia
▨ Spain
■ USA

Figure 21 The colonial possession of the world in 1914

Many millions of western European people spread all over the world settling across the other five continents and specialising in living in far-flung islands. Each act of colonisation brought peoples of different backgrounds and ethnicities together, as white Europeans met, attempted to trade with, lived among and fought with native populations that were ethnically different in every way. White Europeans went to new places and interacted with people with very different ethnic backgrounds. They then brought these 'exotic' people back with them to their home countries.

The creation of diversity in British colonies

The first settlers to Australia arrived in the 1840s; these were mostly convicts who had been deported from Britain. In the early 1870s the promise of land and an escape from poverty meant that many more people chose to leave Britain and travel to the new 'settlement colonies'. Large numbers of people also moved to New Zealand, Canada and South Africa, but often settlement required the newly arrived settlers to take land from the native people; between 1840 and 1870 there were a number of wars fought with the native Maori peoples in New Zealand over land. In 1911, 455,000 people left Britain to move to one of these dominion colonies.

Colonisation is often seen as a way of absorbing and assimilating foreign people into the culture and lifestyle of the imperial country. It involved the physical settlement of people (settlers) from the imperial centre to the colonial periphery. Often this had the impact of destroying the cultural traditions and heritage of an area. Usually colonisation took place to help exploit a particular resource.

Annexation

Sometimes **annexation** is seen as taking political control of a neighbouring country; however, it is a formal act by which a state will take over the sovereignty of another territory. Most annexations do involve one country taking over an adjoining area but this is not always the case. Annexations are usually completed forcibly: militarily, inadvertently through pressure, bribery or politically. Famous examples of annexation include when Germany annexed Austria in 1938 or when Indonesia annexed East Timor in 1975. Sometimes the annexation is taken in an attempt to free an ethnic group from the rule of a less sympathetic state.

In March 2014, Russian President Vladimir Putin announced the annexation of the Crimea two days after a referendum taken in the area on a decision to separate from Ukraine. He said that 'in our hearts we know that Crimea has always been an inalienable part of Russia'. However, the Ukrainian Prime Minster, Arseniy Yatsenyuk, called this 'a robbery on an international scale and one that Kiev will never accept'.

> **Annexation** is when a state takes control of another without permission.

International migration

It would be extremely difficult to measure the extent and the impact of global **international migration**. The UN describes it as 'a global phenomenon that is growing in scope, complexity and impact. Migration is both a cause and effect of broader development processes and an intrinsic feature of our ever globalising world.'

Migration is the single biggest reason why people from different ethnic communities have come into contact with each other. Migration is part of colonisation and annexation as both processes require a movement of people either to settle or to control the area. In most cases the reasons for migration are much more personal and involve a decision to move to a better place for economic or social reasons (pull factors) or a move away from war, oppression or famine (push factors).

Migration is often a necessary process to keep the economic progress of a country on track. The USA has long been noted as a country built on migration; John F. Kennedy called the USA 'a nation of immigrants'. Its history and success at becoming one of the richest, most prosperous and most powerful countries in the world has been built on migration (Figure 22).

> **International migration** is the movement of people from one country to another country.

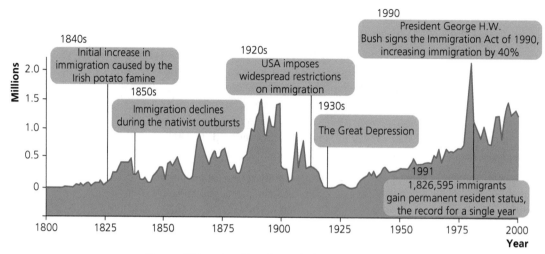

Figure 22 Immigration through the years in the USA

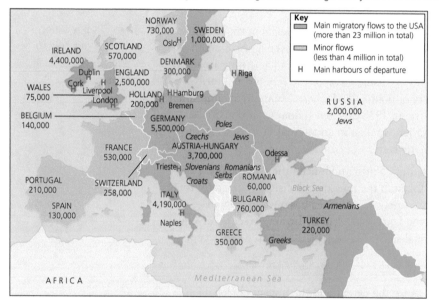

Figure 23 European emigration to the USA (1820–1920)

Exam tip

The processes that create diversity do not fit well into every case study, so you might need to describe only two out of the three in any one answer. However, it is best to have detailed knowledge of all three processes.

Knowledge check 18

Which of the three processes for creating ethnic diversity is the most important?

Case study

The role of processes in creating ethnic diversity in a country

In the UK there has been a long process of political intrigue; decisions made by the different rulers have meant that Britain has long been a diverse and varied nation.

Colonisation

Colonial interactions brought the British people into contact with people of different race, religion, language and nationality. Colonialism required the soldiers/colonists to interact with the native populations in order to trade, farm, exploit resources and to explore the further potential of lands.

The growth of the British Empire continued quickly through the eighteenth and nineteenth centuries: the colonies in America and Canada continued to expand; Captain James Cook made voyages of

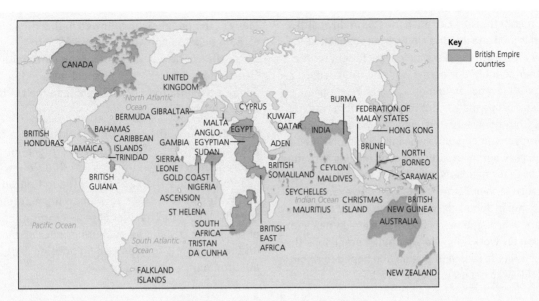

Figure 24 The British Empire in 1899

discovery to Australia, New Zealand and around the Pacific islands. The scramble for Africa led to the UK acquiring colonies that stretched the length of the continent from Egypt in the north to South Africa in the south (Figure 24). At its peak in 1913, the British Empire consisted of 412 million people (23% of the global population) and covered 35.5 million km^2 (24% of the area of the Earth).

Annexation

The expansion of the UK came about as England gradually exerted pressure and invaded its neighbours. Initially England would attempt to set up colonies and seats of power in each country.

The Acts of Union in 1707 (Scotland) and 1801 (Ireland) showed how England was able to annex its neighbours and set up the same level of political control on them. The countries became 'United into One Kingdom by the Name of Great Britain'. This emphasised the political annexation that had been in place since 1603 when King James VI of Scotland inherited the English throne from his cousin, Queen Elizabeth I. It followed the Laws of Wales Acts in 1535 and 1542 which made Wales a full and equal part in the Kingdom of England.

The further Acts of Union in 1800 united the Kingdom of Great Britain and the Kingdom of Ireland to create the United Kingdom of Great Britain and Ireland. These acts further consolidated the political control that England had on its neighbouring peoples. The King of England had been the designated ruler of Ireland since the Crown of Ireland Act in 1542; this was seen by many (including the eighteenth-century Irish Patriot Party) as an Act of Annexation.

International migration

In recent years the main migration stream has been one where people have moved into the UK from the former colonies and empire, but also from European neighbours and further afield.

UK census figures show that the number of people who were born abroad but were living in the UK was very small until the middle of the twentieth century. Some estimate that the foreign-born population of England and Wales had increased to about 7.5 million people (13% of the population) by 2011.

Following the Second World War, many displaced people were looking for new places to settle. Labour shortages in the UK and Europe caused some people to move, for example, 15,700 Polish people settled, mostly due to ties made during the war.

The British Nationality Act in 1948 allowed subjects of the British Empire to live and work in the UK. ➜

This meant that any citizen in the Commonwealth could immigrate into the UK without restriction. Many men from the West Indies had fought for the 'mother country' during the war and returned to lives with few opportunities. Labour shortages across Europe meant that the government turned to its Commonwealth to supply necessary labour. The movement of people was initially seen as being temporary but the reality is that as the workers made new lives for themselves in the UK they were unlikely to return to their country of origin. One example of this came in June 1948 when a ship called the *Empire Windrush* brought nearly 500 men from Jamaica to London for work. Many former servicemen took this opportunity to return to Britain and hoped to rejoin the RAF or to get manufacturing jobs.

Many of the people who followed in this migration stream became known as 'the Windrush generation'. They usually got jobs in cities with British Rail, the NHS or Transport for London. Today, an estimated 172,000 people in the UK have a West Indian-born descent.

In August 2016, the UK Office of National Statistics noted that the latest net migration figures showed an estimated 831,000 Polish-born residents. More than 25% of births in England and Wales in 2015 were to women born outside the UK. Migration has become a common feature of life in the UK and as such its impact on society, on the economy, on jobs and on politics is at an all-time high.

Processes that maintain ethnic diversity

Once diversity has been established in a population it cannot be assumed the original traits that caused the diversity will continue. Often over time people will start to assimilate and integrate into a homogenous community. However, sometimes decisions are made which mean that separate ethnic groups will interact but not assimilate.

Segregation

People can be segregated in many different ways, from housing segregation, racial segregation, religious segregation, residential segregation, occupational segregation or age segregation. Segregation is all about living separate lives. Extreme cases of segregation have taken place in some US cities where black ghettos have been established and there is associated racial violence.

The government policy of apartheid in South Africa meant that people of different colour were forced to live in different parts of cities and rural areas. It led to the relocation of millions of people. The policy was introduced by the National Party in 1948 and attempted to ensure that people of different races would not live side-by-side. Mixed marriage and illegal squatting were banned and separate amenities and services had to be provided for the different races. Education, employment and government became increasingly segregated and there was a huge difference in the way that black people and white people lived in the country. Racial discrimination was deliberately incorporated into the state policy which meant that most black households suffered poor living conditions with fewer opportunities than the white population.

Multiculturalism

The huge amount of migration into the UK following the Second World War has had a massive impact; the UK has become an increasingly ethnic and racially diverse state with different world religions, languages, residential concentrations and social identities.

Exam tip

Make sure that you bring the case study as up to date as possible. What aspects of annexation or migration are currently in the news?

Assimilation is the process where a person's culture comes to resemble that of another group. Full assimilation occurs when new members of a society are difficult to distinguish from the older members of the group.

Integration is the movement of minority groups into the mainstream society.

Segregation is when there is physical separation between different ethnic groups.

Multiculturalism is a political decision to allow people from other countries/nations/ethnicities to continue to display their own cultural identities rather than to assimilate into the majority culture. This means that ethnic groups will remain distinct rather than assimilating into the dominant cultural mainstream beliefs and ideologies.

In 2011, Prime Minister David Cameron made a speech in Munich where he rejected 'the doctrine of state multiculturalism' for having 'encouraged different cultures to live separate lives, apart from each other and apart from the mainstream'.

The Canadian government is often seen as being the model for political awareness of multiculturalism. The Canadian Multiculturalism Act became national policy in 1971. Canadian students are taught about multiculturalism in school, looking at rights and privilege in relation to race, class, gender, religion, disability and sexual orientation.

A society that has achieved an extent of cultural diversity has been influenced by pluralism and might be described as a plural society. Often the main things that divide in a plural society are race, language or religion.

Economic, social and spatial outcomes of ethnic diversity

Minority communities quickly learn to live in relative peace in pluralist or multicultural societies. It is rare for diversity to become a form of conflict; however, often diversity will lead to a consolidation and hunkering down of attitudes which can lead to a breakdown in trust and a lack of social cohesion.

Economic outcomes

The deliberate segregation of peoples can cost money through the building of walls and fences and the cost of security and policing the safety of the different areas. Equally, multiculturalism allows different cultures, languages and religious beliefs to be established. Schools and health care providers will need to have translation facilities. However, diversity can also bring opportunity as new communities help to share their cultural identities and set up ethnic food outlets.

Social outcomes

Diversity can create a deliberate division between people that can be difficult to overcome. The behaviour and attitude of one group towards another group or groups can influence the way that each group interacts with the others. In the majority of cases, the dominant group can discriminate against the other groups and can have economic, political and social advantages that the minority group do not share. Discrimination and extreme segregation can cause a feeling of isolation as people feel that they are leading separate lives.

Spatial outcomes

Often the diversity between groups is manipulated by the dominant group so that there is a clear spatial segregation between the groups. This can be done informally, or on an economic basis (so that the poorer people from a group live in particular parts of the inner city). Usually it is visible through a sense of territoriality where the group will project the shared identity, language and symbols that divide them from the other groups. Sometimes the segregation is reinforced by the government and is made permanent so that continued division is seen as protecting both communities.

Multiculturalism is the existence, acceptance or promotion of multiple cultural traditions in a country or area.

Exam tip

Make sure you have a good understanding of what can maintain ethnic diversity in an area. The main processes are segregation and multiculturalism. Are there more?

Knowledge check 19

What are integration and assimilation in relation to social outcomes of diversity?

Exam tip

Make sure you learn the Belfast case study in detail. You need to focus on both the processes that maintain ethnic diversity *and* the outcomes.

The role of processes in maintaining ethnic diversity and their outcomes for an ethnically diverse city

The history of the city of Belfast has long been dominated with differences and diversity between different groups of people.

Processes maintaining ethnic diversity

Belfast has experienced a huge amount of segregation, discrimination and multiculturalism over recent years.

Segregation

Much of the separation in Belfast is along ethnic lines. Many areas of the city can be clearly identified as either Catholic or Protestant. However, in many ways this is an over-simplification as there is also a division in social class in each of the groups.

Protestants and Catholics did not trust each other, did not communicate or come into contact with each other and lived entirely separate and segregated lives where they went to different churches, different pubs, different schools and different shops. The patchy segregation that was evident on a small scale in Belfast before 1969 was further entrenched as people moved house from mixed communities to ones that were exclusively either Protestant or Catholic. Religious enclaves were established so that people felt safety only with people who were similar to themselves. One of the things that Boal (1969) noted was that in 1967, at one interface zone in West Belfast between the largely Protestant Shankill Road and the largely Roman Catholic Falls Road (Clonard) area, there was a zone where there was a mix of the two communities (Figure 25).

In the months that followed the drawing of this map, the unrest, looting and riots that punctured life across West Belfast meant that there was a consolidation of Protestants back into the Protestant enclaves and of Roman Catholics into the Catholic enclaves. Intimidation caused an estimated 1,400 people per year to move. The ethnic mixing disappeared and peace walls and lines were established to protect and to separate the two groups of people from fighting against each other and to make control more achievable. There are 48 such lines across Belfast today, covering a distance of 21 miles (Figure 26).

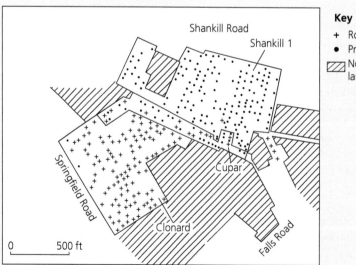

Figure 25 The Shankill/Springfield area of West Belfast (from Boal, 1969)

Table 13 A simplification of the Belfast ethnic diversity/political landscape

	Minority Catholic/ Nationalist community		Majority Protestant/ Unionist community	
Religious affiliation	Usually Roman Catholic	Usually Roman Catholic	Usually Protestant	Usually Protestant
Political affiliation	Republican (Sinn Fein)	Nationalist (SDLP)	Unionist (Ulster Unionist)	Loyalist (DUP/PUP)
Socio-economic group	Working class	Middle class	Middle class	Working class
Paramilitary groups involved	■ Provisional IRA ■ Official IRA ■ Irish National Liberation Army (INLA)		■ Ulster Volunteer Force (UVF) ■ Ulster Defence Association (UDA) ■ Red Hand Commando (RHC)	

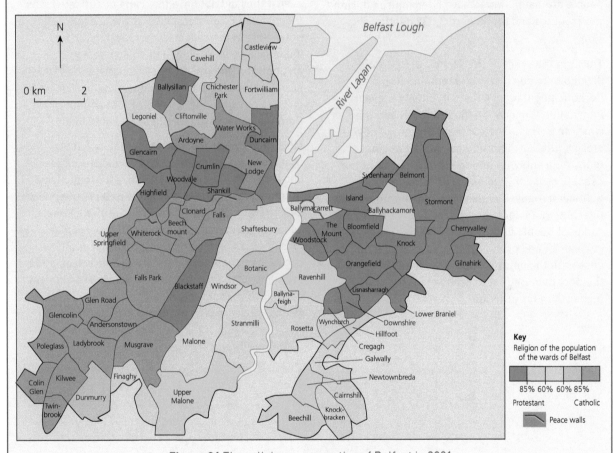

Figure 26 The religious segregation of Belfast in 2001

Often this religious segregation led to a huge amount of discrimination across the city. The Catholic minority population felt that they did not have equal representation politically, fairly drawn electoral districts, fair access to jobs and housing or the right to speak up about this. This created a sense of alienation which created a division in opinion with the Protestant/Loyalist community.

The high levels of segregation made it much easier for each ethnic group to display their religious and political affiliations. This was done to mark their territory but also as a form of intimidation to those on the periphery. Gable walls were painted with political slogans and murals, flags were draped from lampposts and kerb stones were painted in the affiliated colours.

The segregation in the city was further concentrated due to the Troubles, the period of violence and unrest that existed as conflict in the wider NI community from the late 1960s to 1998. The violence is often described as a guerrilla war. The main participants in the Troubles were the republican paramilitaries (IRA and INLA); loyalist paramilitaries (UVF and UDA); and the British state security forces: the British Army and the Royal Ulster Constabulary (RUC). In total over 47,500 people are estimated to have been injured during the Troubles and there were 3,532 deaths.

Multiculturalism

Through the years of the Troubles, Belfast was divided between the Protestant and Roman Catholic population. Other UK cities received international migration though few made their way into Belfast. The constant stream of news stories filled with bombings and killings put many migrants off coming into the city until the ceasefires from 1996 onwards and the fresh political initiatives. Since then, huge amounts of international migration have brought a diverse range of people into Belfast. This has created its own social and economic issues as Belfast tries to cope with the influx of migrants; it has also allowed the people of Belfast to realise that the world is bigger than they thought.

Outcomes of ethnic diversity

The ethnic diversity in Belfast from the late 1960s to 1998 (and beyond) caused a variety of different outcomes that made sure that the city of Belfast was unable to develop in the same way as other cities on the UK mainland.

Economic outcomes

Until the 1970s, employment in NI was highly segregated. Protestants were able to get better jobs even if they did not have the same qualifications as their Catholic counterparts. This was especially noticeable in the civil service, manufacturing industries such as shipbuilding and heavy engineering and the police. Throughout the 1990s a series of laws was introduced to stop discrimination on religious grounds. The Fair Employment Commission was established to uphold this.

Much money was lost due to the violence that occurred throughout the city. Many shops, offices and factories were blown up and lost productivity. Unemployment remained high in Belfast, often above 24% (compared to the UK average of 13%). People became reluctant to venture into the city centre to do their shopping and the city was on unofficial curfew each night from 6pm as people retreated to the safety of their own homes. This also resulted in the loss of tourism revenue. The number of people visiting NI decreased dramatically (Figure 27).

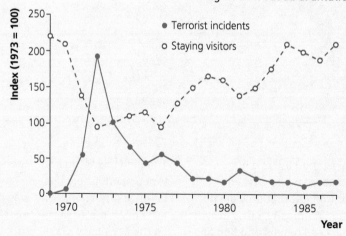

Figure 27 The impact of terrorist incidents on tourism in Belfast

Table 14 Breakdown of deaths due to the Troubles in NI (1969–1998)

Civilians	UK security services (1049)	Republican paramilitaries (368)	Loyalist paramilitaries (162)
Civilians/ non-combatants (1,841)	British Army 705	PIRA 291	UDA 91
	RUC 301	INLA 39	UVF 62
Irish Security services (11)	Prison service 24	OIRA 27	RHC 4
	Other UK services 19	Other 11	Other 5

Social outcomes

The ethnic diversity in Belfast that resulted in the Troubles caused a huge number of deaths (Table 14); over 3,500 people were killed and nearly 50,000 were injured as a result. This had a massive impact on NI society at large. People from the different communities failed to trust people from the 'other' community. Many felt that violence had become dehumanised and death was commonplace.

The NI education system is heavily segregated. The state schools in NI are usually seen as being Protestant with the majority of Catholic children attending schools that are maintained by the Catholic Church. This has meant that many young people in NI are unlikely to speak to or to be friends with someone from the other religious group until they attend university or get a job. The integrated education movement has attempted to set up both primary and secondary schools where children from all backgrounds can be educated together and help heal division in the country.

Initially public/social housing was also extremely segregated between the two communities. The amount of residential segregation increased from about 60% in 1969 to 93% in 2004. However, one outcome has been that there is now equality in the distribution of social housing across Belfast and someone in need of social housing is less likely to be discriminated against due to their religious beliefs.

Spatial outcomes

One of the most obvious outcomes of the diversity that has rocked Belfast over the last 40 years has been segregation. Fred Boal wrote that 'Segregation is usually measured on the basis of the spatial distribution of individuals or households belonging to a particular economic or ethnic group.' The Belfast society of the 1960s was seen to be both polarised and ranked. The polarised communities had no interaction between their distinctive component groups.

The conflict across Belfast made sure that there was a deep and physical separation of people. Small, homogenous groups lived across the city and had no contact with the 'other side'. In fact, the residential concentration of groups often took on tribal proportions so that people felt that they lived in their own little enclaves. In many ways this has been one aspect of life that has been the slowest to change in Belfast. The segregated areas and peace lines remain across the city. However, there are a number of organisations committed to helping to break down the walls of sectarianism in Belfast. In 2013, the Northern Ireland Executive resolved to remove the peace lines across the city by mutual consent by 2023.

Summary

- Ethnic diversity is a common feature in countries and nations. The main processes that create diversity are colonisation, annexation and international migration.
- The differences between people can continue for many years and are maintained through the processes of segregation and multiculturalism.
- Often the maintenance of ethnic diversity over a long period of time can bring economic, social and spatial outcomes.

Ethnic conflict

Many feel that conflict is an inevitable outcome of ethnic diversity when the level of division has reached a certain point. Conflict can take a number of forms, from a strike or peaceful protest to a war. Conflict occurs when people have a disagreement in respect to differences of interest, need or point of view.

Causes of ethnic conflict

Across the world there are many different causes where disagreements along ethnic lines will mutate into conflict. In most cases there is more than one specific cause of the conflict; there can be a number of different causes.

Territorial disputes

The struggle for territory is one of the biggest causes of conflict.

The CIA *World Factbook* online keeps an up-to-date list of disputes; it lists border disputes between Sudan and South Sudan (in the Abyei region), between the Republic of China and the People's Republic of China, and between North Korea and South Korea. Argentina is still keen to assert a claim to the UK-administered Falkland Islands. The Kashmir region of India remains one of the world's largest and most militarised territorial disputes as parts of the area are ruled by China (Aksai Chin), India (Jammu and Kashmir) and Pakistan (Azad Kashmir and Northern Areas).

Historical animosities

Many conflicts around the world are embedded in historical disputes and animosities. Conflict is further fuelled as younger generations are told stories to increase the hatred for the other side. For example, there have long been deep divisions between Japan and China, between Korea and Japan, and between Spain and people from North Africa. Closer to home, in NI historical animosities have long divided the Catholic and Protestant people. Sectarian celebrations such as the annual marching season, including 12 July, bring past battles and issues back to the fore and create further division between neighbours, often leading to tension, stand-offs and riots.

Racism

Racism is not a new phenomenon; people have been progressively discriminated against because they look different from others for a very long time. In the USA there has been a long history of racial discrimination. The Black Codes Laws were used in southern USA to restrict the civil rights and liberties of African–Americans. These were followed by the Jim Crow Laws, which persisted until the 1960s. Both sets of laws created segregation between African–Americans and European Americans, which created big differences in economic opportunities (job discrimination), education, social disadvantages, housing, bank lending and labour union practices.

Racism was also seen as being the main driving force behind the slave trade between the fifteenth and nineteenth centuries. A slavery triangle existed as Europeans from the 'old world' would export goods to Africa. Slaves were captured in the west of Africa and taken to the 'new world' (North and South America) to work on European-owned plantations. Finally, goods from the new world (cotton, sugar, tobacco and molasses) were exported back to Europe.

Knowledge check 20

What is the definition of ethnic conflict?

Racism is the discrimination and prejudice towards people based on their race or ethnicity.

Sectarianism

In NI, the Troubles are seen by many as being a **sectarian** conflict on religious/political lines. Both groups ascribe to being part of the wider Christian religion and the main differences between the Catholics and Protestants are theological rather than ideological. As Protestants in NI became increasingly dominant following the partition of Ireland in 1921, they were able to use their power and influence to keep the Catholic community in a minority position.

Recently within Islam there has been conflict between the Sunnis and the Shias. Shi'ites disagree with the Sunnis as they refuse to accept the first Caliph as Ali ibn Abi Talib and do not see him as being infallible and divinely guided. In response, the Sunni leaders consider the Shias to be heretics and apostates. The main countries with Shia Islam are Iran, Iraq, Azerbaijan and Bahrain. Sunni Muslims make up about 80% of the Islamic population and are found mostly across North Africa, Saudi Arabia and Turkey.

Sectarianism is a form of discrimination that can happen within an ethnic group. The most common feature of sectarianism is when bigotry or hatred build up between two sects with the same religion.

Cultural conflicts

Cultural conflicts exist when there is conflict between people who have different cultural values and beliefs. Often they are extremely difficult to resolve due to the differences in beliefs. Ethnic cleansing is often seen as an extreme example of cultural conflict.

Cultural conflicts can exist when there is discrimination and issues between groups from two different religions. In Nigeria there have long been disagreements and conflict between the Muslim-dominated North and the Christian-dominated South. Over the last 20 years there has been a series of riots in Abuja (2000) and Jos (2001). In 2002, 200 people died in Kaduna and in 2005, 100 people died in Onitsha.

In 1994, long-standing cultural conflict between the Hutu and Tutsi tribes in Rwanda led to ethnic cleansing (genocide). The Hutu majority government killed nearly 1 million people in 100 days from 7 April to mid-July; these people made up about 70% of the Tutsi and 20% of the Rwandan total population. A further 2 million Rwandans were displaced by the violence and became refugees.

Human rights abuses

Following the Second World War, the human rights movement was determined to ensure that the atrocities of the Holocaust would never be repeated. The UN adopted the Universal Declaration of Human Rights in 1948, which consists of 30 Articles that became international law in 1976.

Abuses or violations of these human rights take place when states abuse, ignore or deny these basic rights. There are many non-government organisations that monitor the adoption of these rights, such as Amnesty International, which was founded in 1961 to draw attention to human rights abuses.

Between 1948 and 1991, the apartheid system embedded racial segregation and discrimination in South Africa. Apartheid allowed white-minority rule in the country while the black majority were segregated in their use of public facilities, social events, access to housing and employment.

Discrimination

Discrimination can maintain ethnic diversity but as it becomes further embedded, the reasons for the discrimination can lead to the diversity giving way to conflict. Each of the causes of ethnic conflict has some form of discrimination at its root. Minor discrimination activity might involve age, disability, employment, language, gender, nationality or social class, but these will rarely result in conflict. The main forms of discrimination that lead to conflict are in relation to race (or ethnicity) and religious beliefs.

Modern-day racial discrimination often indicates that some form of ethnic penalty exists in society. In the USA, 40% of prison inmates are African–American (compared to 13% of the whole US population), and 19% are Hispanic (and make up 19% of the whole US population). In Canada the aboriginal people (First Nations, Metis and Inuit) account for 23% of the prison population (and make up only 4% of the population). In New Zealand, 50% of the prison population are Aboriginal Maori (out of 15% of the population).

Discrimination is the consideration of making a distinction in favour or against a person due to the ethnic group that they belong to.

Nature of ethnic conflict

The nature of any potential conflict depends on how divisive the issue is and how strongly held the particular beliefs are.

Civil disobedience

Civil disobedience is often regarded as being a symbolic violation of the law. It usually begins as non-violent resistance, but can sometimes become violent unintentionally.

Mahatma Gandhi became famous for leading Indians in challenging the British salt tax in 1930 and later called for the British to leave India in 1942. Gandhi tried to ensure that all actions were non-violent but some sections of the Quit India Movement took part in small-scale violence and there was widespread rioting with the British losing control of some parts of the country.

In the USA, civil rights leader Martin Luther King Jr also argued that non-violent civil disobedience should be used in order to combat racial inequality. The March on Washington for Jobs and Freedom took place on 28 August 1963. Specific demands were made to end racial segregation in public schools, promote meaningful civil rights legislation, prohibit racial discrimination, protect the civil rights of workers from police brutality and create a $2 minimum wage. Over 250,000 people attended the event and King delivered his famous 'I have a dream' speech.

Exam tip

There are many different causes of ethnic conflict. Ensure you understand them all and know examples of where they have taken place.

Civil disobedience is the active refusal to obey certain laws or commands by a government.

Terrorism

Often when civil disobedience is not having a desired impact, groups will look to use violence (terror or fear) in order to achieve their objectives. This is called **terrorism**. Many ethnic conflicts do not reach full-scale civil war. The UN Comprehensive Convention on International Terrorism notes that terrorism is 'Criminal acts intended or calculated to provoke a state of terror in the public, a group of persons or particular persons for political purposes that are in any circumstance unjustifiable, whatever the considerations of a political, philosophical, ideological, racial, ethnic, religious or any other nature that may be invoked to justify them.'

Terrorism involves an armed campaign with the terrorist being part of an illegal guerrilla organisation.

Terrorism is considered as a major threat to society and as a war crime when used to target civilians. The Global Terrorism Database notes that there have been more than 61,000 incidents of non-state terrorism claiming over 140,000 lives between 2000 and 2014.

Civil war

The aim of a **civil war** might be for one group to take control of the country at the expense of the other. The war may be a high-intensity conflict involving regular armed forces in sustained, organised and large-scale activity. From the end of the Second World War until 2000, there were an estimated 138 civil wars resulting in the deaths of around 25 million people and the forced migration of millions more. Political scientists note that a civil war must have more than 1,000 casualties per year.

The American Civil War took place between 1861 and 1865. In 1861, 11 of the southern slave states declared their secession from the USA and formed the Confederate States of America. The Union (or US) was made up of the 23 free states and 5 border states. The civil war was based on the issue of slavery. During four years of fighting nearly 750,000 soldiers were killed, much of the south's infrastructure was destroyed, the Confederacy collapsed and slavery was abolished.

A **civil war** is a war that takes place between organised groups in the same state or country.

Outcomes of ethnic conflict

It is rare for ethnic conflict to fizzle out but usually the nature of violence used in the conflict will create a series of far-reaching consequences that people have to deal with for many years to come. The conflict is by its very nature a further divisive factor that can continue to entrench opinion and diversity, making it increasingly difficult to find a resolution to the violence.

Social and economic impacts

Conflict usually leads to the destruction of economic and social infrastructure and the environment. Economic damage can range from the direct destruction of production and other facilities (like tourism and airports) to the disincentive of outside investment caused by risks associated with the conflict. Similarly, the social infrastructure — schools, hospitals, churches and shops — can be damaged or suffer from lack of investment.

Conflict is destructive to the environment and can cause major long-term issues. The costs of repairing destruction caused by violence take finances away from development goals. Many areas of conflict in recent years continue to be blighted with unexploded ordnance and landmines.

Some conflicts have actually meant that countries have experienced reverse development. The World Bank noted that some African countries experienced a sharp decline in the GDP per capita following civil war/conflict (Table 15).

In some ways the people who are impacted by conflict has changed. At the start of the twentieth century nearly 90% of all conflict fatalities were soldiers, but by 1990 nearly 90% of fatalities were civilians. This emphasises how the nature of conflict has often moved quickly into devastating civil war where access to civilians is common.

Table 15 The GDP per capita before and following conflict

	Before conflict (in US$)	Following conflict (in US$)
Rwanda	306	181
Burundi	207	143
Democratic Republic of Congo	122	103
Guinea-Bissau	240	176
Sierra Leone	214	150

Conflict in the Darfur areas of western Sudan led to around 2.7 million people becoming refugees. Many of these people lived in camps near Darfur's main towns. Around 200,000 also fled to Chad where they lived in camps near the border. An estimated 300,000 people were killed in the conflict as a result of war, famine and disease. Many of the refugee camps continued to suffer violent attacks. The murder of local farmers also meant that there was insufficient food for the people.

Territorial division

Sometimes the only way to ensure that conflicts are resolved is to divide a territory into separate and distinct countries. Following independence from Great Britain in 1947, India and Pakistan became two separate countries. Although this did create two very different cultural identities, it did not completely work as some territories were, and are still to this day, disputed territories which can cause concern and continue to cause minor conflict (the Punjab/Kashmir).

Autonomy

Quebec and Nunavut in Canada are examples of places with a high degree of **autonomy** in Canada. They have a degree of self-rule but are not independent. Northern Ireland was given autonomy as part of the 1998 Good Friday Agreement (called devolution). Hong Kong is an area that has been given autonomy as a special administrative region in China, which was made in agreement as part of the Sino-British Joint Declaration of 1984. Easter Island in the Pacific was made a special territory by Chile in 2007. Often the reason autonomy is offered is to make sure that a specific area will get special treatment, which might stop it from seeking full independence from the rest of country, which might involve some aspect of conflict.

Autonomy is when an area has a level of self-governance.

Ethnic cleansing

In some conflicts the violence used in the civil war will involve a systematic and forced removal of one particular ethnic or religious group. The forces applied might attempt a forced migration but usually will instigate intimidation, mass murder and genocide. In many cases of ethnic cleansing there will be some attempt to remove evidence of the violence committed.

The phrase has long been used across history but the Nazi Germany policy during the Holocaust aimed at ensuring that Europe was 'cleansed of all Jews'. Organisations such as Genocide Watch have been set up to monitor, stop, prevent and predict acts of genocide globally.

International intervention

Ethnic conflicts are sometimes subject to outside intervention from a neighbouring country, regional power, world power, or even the UN. Any effective outside intervention will depend on the strategic interests of the outside power. For example, in the conflict between the Tutsis and the Hutus in Rwanda and Burundi in the 1990s the outside world displayed little willingness to become involved, other than in caring for the refugees. In Yugoslavia, when the Muslims were in conflict with the Serbs in Kosovo, the western powers acted much quicker with more military might.

The UN Charter notes that any case of serious human rights abuse is a case for international concern and intervention. However, many people in the LEDC world note that the countries that are quick to seek international interventions are countries like Britain, France and Canada: all states that might have a strong desire to export their cultural tendencies — liberal democracy and free-market capitalism.

Peace processes

The phrase 'peace processes' sounds as if they should happen following conflict, but the phrase is often used to describe attempts to try to bring the conflict to a close. In NI the term was used before the ceasefires of the mid 1990s until the embedding of the power-sharing executive in the NI Assembly. The process for peace in NI has not been smooth or continuous, but it has been more successful and more stable than other conflicts.

> **Exam tip**
>
> Detailed knowledge of the different outcomes of ethnic conflict is required, especially in ensuring you show depth of knowledge in your case study.

Case study

The role of the processes of ethnic conflict that have affected a nation

Causes of ethnic conflict in Cyprus

Historical animosities

Throughout history the island of Cyprus has largely been associated with Greece. From 1570 it was conquered by the Ottoman Empire which started three centuries of rule over Cyprus. By the start of the nineteenth century the ethnic Greeks on the island were keen to unite Cyprus with Greece. However, the UK took administrative control of the island in 1878.

Throughout the early years of the twentieth century there was ever-increasing tension between Greeks and Turks. The Greek Genocide between 1913 and 1922 left up to 750,000 Greeks dead. The Greco-Turkish war lasted from 1919 to 1922.

Although Britain declared Cyprus to be a Crown colony in 1925, there was still a movement for Cyprus to join with Greece. Turkish Cypriots consistently opposed any union of this type and in 1931 there was an open revolution which resulted in the deaths of six civilians and the burning of the British Government House in Nicosia.

Territorial disputes (including civil disobedience and terrorism)

Following the Second World War, most of the Greek Cypriots were still adamant to pursue the union with Greece. In 1950 Archbishop Makarios III noted that he would not rest until union (*enosis*) with Greece was reached.

In 1955 the National Organisation of Cypriot Fighters (EOKA) was founded and they began an armed campaign against the British. This led to the death of 387 British servicemen. Tensions rose between the two Greek and Turkish Cypriot communities. In 1957 the Turkish Resistance Organisation (TMT) was formed to protect the Turks and to ensure that the island would be divided in the event of any union with Greece.

→

Cultural conflicts (including civil war)

By 1958 the island was on the verge of civil war. However, the Zurich–London agreements between the Greek Cypriots, Turkish Cypriots and British were able to establish a framework for self-government. Although the three treaties were generally accepted, many on both the Greek and Cypriot sides were not content because they had not achieved everything that they had hoped for. Cyprus achieved independence on 16 August 1960. A power-sharing governmental system was put in place but it was quickly stifled by the complexities of its organisation. By December 1963 a constitutional crisis led to the Bloody Christmas where fighting erupted between the main communities in Nicosia. The violence spread quickly across the island. In all, 270 Turkish Cypriot mosques and shrines were vandalised and around 133 Turkish Cypriots were killed over a ten-day period.

Over the next few years 364 Turkish Cypriots and 174 Greek Cypriots were killed and 18,667 Turkish Cypriots were forced to abandon their homes. From 1964 to 1974 there were a number of international efforts by the UN and the USA to bring some sort of resolution to the political impasse.

In the early 1970s Archbishop Makarios was losing control and was overthrown by Nikos Sampson who was intent on making sure that Greece took full control of Cyprus. On 20 July 1974 the Turkish Prime Minister Bulent Ecevit ordered a military invasion of Cyprus.

Operation Atilla was launched and the Turkish forces invaded and captured 3% of the island before a ceasefire was declared. A further wave of forces landed between 14 and 16 August and took nearly 40% of the island. The ceasefire line from August 1974 became the UN buffer zone and is known as the green line, leaving Cyprus in a state of partition (Figure 28).

The impact of the war meant that:

- The Turkish now occupied 36% of Cyprus.
- The Cypriot and Greek military rulers lost power on 23 July 1974.
- An autonomous Turkish Cypriot Government was set up and eventually became the Turkish Republic of Northern Cyprus.
- Between 150,000 and 200,000 Greek Cypriots were displaced from the north in place of between 40,000 and 65,000 Turkish Cypriots in the south.
- The conflict also caused the deaths of 1,641 Turkish Cypriots and around 24,000 Greek Cypriots.

In 1983 the Turkish Republic of Northern Cyprus (TRNC) declared independence, but Turkey is the only country that recognises this. The remainder of the wider international community class the northern area as Turkish-occupied territory of the Republic of Cyprus. When Cyprus became a member of the EU this annexation of territory was declared illegal.

Human rights abuses

The conflict sparked a few examples of human rights abuses.

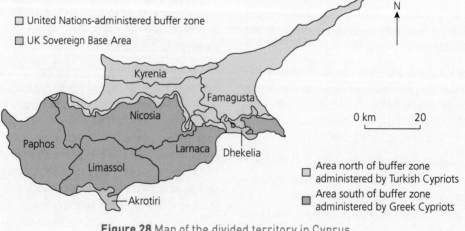

Figure 28 Map of the divided territory in Cyprus

- The Martatha, Santalaris and Aloda massacre by the EOKA B left 126 dead on 14 August 1974 and the UN branded this as a crime against humanity.
- Another massacre of 36 people was carried out in a village near Limassol by Greek Cypriot forces.
- In Limassol the Turkish area was burned, women were raped and children shot.

Nature of ethnic conflict in Cyprus

In the twentieth century, Cyprus experienced all three aspects that make up the nature of ethnic conflict. The rumblings of discontent began at first with rioting (civil disobedience). This quickly led to each side of the conflict looking to escalate things and take further action (terrorism) in order to achieve their specific demands of alliance with either Greece or Turkey. Eventually, in 1974, the Turkish forces invaded in what became a short-lived conflict/civil war but one which would have far-reaching consequences.

Outcomes of ethnic conflict in Cyprus

Social and economic impacts

As a result of the conflict, many parts of Cyprus were completely destroyed. Infrastructure, roads and airports were ruined. The capital city Nicosia/Lefkosha was in ruins. Varosha, the main tourist area including Famagusta and Gazi Magusa, was left empty. Much of the Cypriot cultural heritage was destroyed or stolen and taken off the island to be sold to wealthy collectors.

It took many years for the tourism industry on the island to recover (and only in the south). The UNHCR continued to give assistance to displaced people right up until 1999. Families had been wrecked; family members had been killed, injured or displaced as a result of the conflict. Thousands of people were living in totally different places from where they used to live. Throughout the early 1980s and 1990s there was a big difference in the GDP between the two areas; the Greek Cypriot GDP was $14,000 per year whereas the equivalent in the Turkish Cypriot side was $4,000.

Territorial division

The most obvious aspect of the conflict in Cyprus has been the separation of the country into two autonomous states. The green line continues to divide the country and the same three interest groups continue to hold territory in the country (Greek Cypriot, Turkish Cypriot and British). For many years the UN sent peacekeepers to keep an uneasy peace between the two communities. Nicosia was a divided capital with a wall separating the two countries.

Autonomy

The Republic of Cyprus was widely accepted as a separate, autonomous state. Even though the island joined the EU in 2004, the laws and rights are only applied to the southern areas of the island.

Ethnic cleansing

Although there were examples of ethnic cleansing noted on both sides during the short-lived conflict, the establishment of a permanent barrier between the two countries prevented further conflict and opportunities for violence against each other.

International intervention

There have long been opportunities and examples of outside agencies becoming involved in this conflict. The UN, the British government and the USA have been actively involved for many years. The UN has led talks on the status of Cyprus since 1999. The Annan Plan for Cyprus was developed in 2004 to attempt to create the United Republic of Cyprus which would be a federation of the two states. It was put to a referendum in 2004 but the proposal was only accepted by 65% of Turkish Cypriots and 24% of Greek Cypriots.

Peace processes

Following a 30-year ban on crossing the green line, the Turkish Cypriot administration allowed travel across the line in April 2003 at the Ledra Street crossing in Nicosia (Figure 29). Members of both communities are now free to cross the buffer zones at some selected checkpoints.

In the 2014 Cyprus talks the Greek and Turkish Cypriot leaders made a joint declaration listing some of the preconditions for talks. Very little further action has resulted in recent years and many feel that the peace process is at an impasse but that some form of federated united Cyprus will take shape at some point in the future.

→

Figure 29 The Ledra Steet crossing in Nicosia (taken in 1992)

Summary

- Ethnic conflict frequently occurs around the world. There are a number of different causes that can start conflict between groups of people: territorial disputes, historical animosities, racism, sectarianism, cultural conflicts, human rights abuses, discrimination.
- The nature of ethnic conflict can be seen through civil disobedience, terrorism and civil war.
- There are a number of outcomes of ethnic conflict: social and economic impacts, territorial division, autonomy, ethnic cleansing, international intervention, peace process.

■ Option D Tourism

The changing nature of tourism

How mass tourism has developed into a global industry

The United Nations World Tourism Organization (UNWTO) notes that, 'Tourism is defined by the activities of persons identified as visitors. A visitor is someone who is making a visit to a main destination outside his/her usual environment for less than a year for any main purpose [including] holidays, leisure and recreation, business, health, education or other purposes.' (2010)

Tourism is when people visit another region or area for a period of at least 24 hours. It can be international or within the country of the traveller.

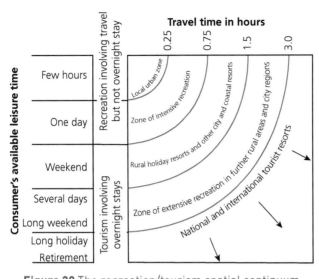

Figure 30 The recreation/tourism spatial continuum

Mass tourism as a global industry

Mass tourism is the most popular form of holiday as everything is packaged together so that the visitor can combine accommodation, local transfers and flights, saving money and making the holiday more affordable.

Over the last 60–70 years, tourism has become increasingly accessible to all people. Up to that point, it was a luxury that only the very wealthy could afford. In the 1950s the cost of a flight was unbelievably expensive and the average person could not afford it. Today, the number of holidays that people plan to go on both domestically (at home) and internationally is at the highest level that it has ever been. In the last few years the number of global international tourist arrivals has topped 1 billion (Figure 31), compared to 25 million in 1950.

Mass tourism usually describes the practice where many tens of thousands of people go to the same resort at the same time of year.

Exam tip

Make sure you are aware of how tourism has developed as a global industry over the last 100 years.

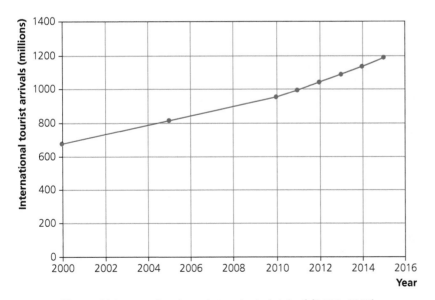

Figure 31 International tourist arrivals (global) (2000–2015)

The global share of these international tourists (Figure 32) shows that a majority of tourists (over 607 million) arrived in Europe. Over 127 million arrived in North America. Far fewer visitors visited South America, Asia and Africa. In 2015 the 1,186 million international tourist arrivals brought $1,260 billion in international tourism receipts (Table 16). The global tourism industry now represents 10% of the global GDP and accounts for 1 in every 11 jobs. The level of domestic tourism (within countries) is estimated to be between 5 and 6 billion people.

Knowledge check 21

What is the difference between tourism and mass tourism?

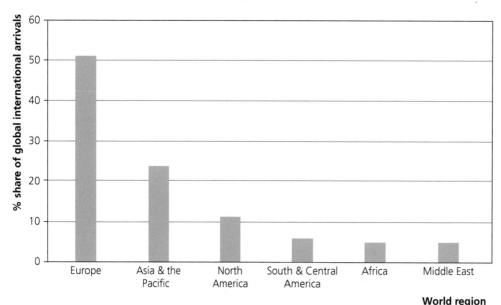

Figure 32 The global share of international tourist arrivals, 2015

Table 16 International tourist arrivals and receipts, top ten countries

Rank	International arrivals (millions)	% change from previous year	Tourism receipts in US$ (billions)
1 France	84.5	0.9	45.9
2 USA	77.5	3.3	204.5
3 Spain	68.2	5.0	56.5
4 China	56.9	2.3	114.1
5 Italy	50.7	4.4	39.4
6 Turkey	39.5	−0.8	26.6
7 Germany	35.0	6.0	36.9
8 UK	34.4	5.6	45.5
9 Mexico	32.1	9.4	17.7
10 Russia	31.3	5.0	8.4

Developments in transport

The development of technology and transportation has helped the tourist industry become more mainstream. As infrastructure improved, journey times from one place to another reduced. The introduction of railways allowed people to travel from one place to another in relative comfort at a reasonable cost. Cruise liners, passenger ferries and planes continued to enable people to start considering visiting more distant places. The rise of the motor car (following the Second World War) meant that an ever-increasing number of people had access to their own means of transport that would allow them to explore their own country and beyond. The development of motorways and the building of new regional airports made travel easier. The time that it took to travel from one place to another lessened, so that it felt as if places were closer and more accessible than at any other time.

Increase in disposable incomes

The work pattern that many people now enjoy helps to facilitate time and money that can be used for holidays. Working conditions in the UK have gradually improved (Figure 33). Fifty years ago, workers did not have the same amount of time off as today; they might have been allowed one week and maybe travelled to a local seaside tourist resort like Portrush or Newcastle for a week of full board lodging in a guest house.

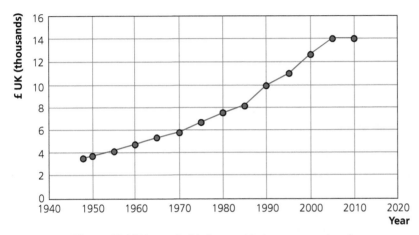

Figure 33 UK household disposable income per head

Workers in the UK today are entitled to 5.6 weeks' paid holiday per year (this is known as the statutory leave entitlement). In addition, workers today are paid much better wages, which means that they have more disposable income to spend on leisure. Many people are able to afford multiple holidays and retired people are often able to afford a few foreign holidays each year.

Rise of package holidays

The first Thomas Cook overseas tour was to the USA in 1855.

The resorts in the Mediterranean became popular in the 1960s as they were marketed as having year-round sunshine. The fact that they have much better weather than the British winter (and summer) meant that package holidays to the hotter climates in Spain became the norm.

The number of British people taking package holidays abroad reached 15.9 million in 2014.

Impact of internet access

Computers have revolutionised the travel and tourism industry. They have enabled air traffic control procedures to become increasingly complex, allow aircraft to fly on autopilot and have automated baggage procedures. Universal internet access now lets people make travel arrangements quickly. There is less need to book appointments with a travel agent; people can quickly scan through multiple holiday options on their computer or smartphone and make bookings themselves. Many of the booking systems are based on complicated central reservation systems. The typical user can quickly use an internet search to check if rooms are available at any hotel around the world and find out the price instantly.

Travel agents have been on the decrease in recent years as people take more responsibility for booking their own travel arrangements. Tour operators can also maximise sales by offering special deals to promote unsold holidays in the last few days before departure.

Positive social and economic impacts of tourism

Tourism can be the catalyst that inspires a place to grow from being a small fishing village into a huge, high-rise city catering for many millions of visitors at a time. Tourism can single-handedly transform the economic and social conditions of any place.

Positive social impacts

- Cultural exchange can be stimulated due to the broadening of horizons. There will also be reduced prejudice between visitors and a local population. The culture of indigenous peoples might be protected due to tourist interest in them, for example, in Nairobi, Kenya a museum highlighting the tribal history of the country called the Bomas of Kenya can be visited.
- Tourism often allows an enhanced role for women in the society. Their status will be increased as they are involved in providing for the needs of tourists.
- Education will be improved directly through better funding (due to increased income from tourists paying more taxes) and indirectly through the contact that students and their families have with international travellers.
- Travel to other countries will be encouraged. There tends to be more mobility and social integration in countries where tourism is growing. Many global hotel chains offer opportunities for staff to go on placement to other hotels around the world.
- Many of the local services and infrastructure will be improved in order to cater for the needs of the tourist (e.g. electricity, health care, transport facilities, roads and airports) plus there will be a greater range of shops, restaurants and leisure services available.

Positive economic impacts

- Tourism in most cases will increase GDP directly and increase the multiplier effect, which brings more and more 'hard currency' into the country.
- The taxes paid by tourists arriving into the country, paying for accommodation and services, will increase the revenue collected by the government, which means that it will have more money to pay for improvements for the population as a whole.
- Tourists coming from rich countries like the USA, the UK and the EU will bring foreign currency with them that is converted into a local currency, giving the banks in the country some hard currency that can help foreign exchange earnings.
- The development of tourism in a new area usually brings foreign investment as hotel chains, global food outlets and leisure companies aim to get their own piece of any expansion.

> **Exam tip**
>
> Make sure that you are clear on the role that each new development has played in the growth of tourism.

> **Exam tip**
>
> Ensure that you understand the positive social and economic impacts that tourism brings to people.

- Tourism provides well-paid, seasonal and permanent jobs that can help to improve the amount of money that an unskilled worker might be able to achieve.
- The influx of tourists into an area will allow the opportunity to make improvements to the local infrastructure that will provide benefits to the local people and tourists alike. Airport facilities will be upgraded, roads and transport links will be developed, security will be improved across cities and at sites of particular interest, new hospital and health care facilities will be built to take care of any visitors who are ill.
- As more money is coming into the country (and staying in the country), this means that the balance of trade (between imports and exports) is improved.

Knowledge check 22

What is the difference between a social and economic impact on tourism?

Changes in tourist demands and resorts over time

People with more disposable income are prepared to spend more money to maximise their experience; they might want to go trekking in the Himalayas or Machu Picchu or walk with penguins in Antarctica. Some will want to ensure that their holiday does not make a big environmental impact so they will try to book an ecotourism holiday to places like Costa Rica or Belize.

The pleasure periphery

Tourists have increasingly held a desire for stranger, further and more exotic holidays. Passport tourism means that people want to go to places just to say that they have been there (and maybe get a stamp in their passport). The main idea behind this concept is that over time people have been able to travel further and further. Increases in prosperity and technology have enabled this movement, meaning that there are few places left on the planet that are not reachable within 24 hours of travel.

Knowledge check 23

What is a definition of the 'pleasure periphery'?

The Butler model

The aim of the Butler model (Figure 34) is to look at the way that tourist resorts grow and develop over time. The ever-increasing reach of tourism means tourists are always on the lookout for the next big thing. Trends and fashions in tourism can change quickly and different countries need to ensure that they remain ahead of the trend if they want to ensure that they maximise their income from the industry.

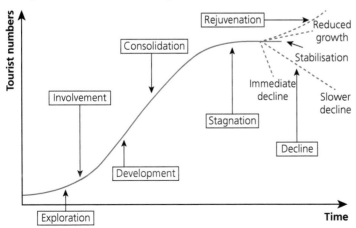

Figure 34 The Butler model

The Butler model has six stages:

1 **Exploration**: Small numbers of early adopters visit an area as they are attracted to the unique natural/cultural/environmental features of the area. There will be few facilities for the visitors, for example, Canadian Arctic or Chile.

2 **Involvement**: Over time the resort starts to grow in popularity and local residents attempt to provide some basic facilities for the tourists. The number of visitors will increase and a tourist season might develop, for example, Mexico.

3 **Development**: Tourist numbers are now expanding rapidly, leading to mass tourism. Package holiday deals are organised by external operators. External organisations provide modern tourist facilities. Tensions will start to build between the locals and tourists. Some labour will be taken up by experienced migrants, for example, Florida or Kenyan safari resorts.

4 **Consolidation**: The number of tourists continues to increase. Tourism is quickly becoming one of the dominant industries in the area. There are more tourists than residents. The resort will no longer be considered as fashionable and prices will be reduced to encourage different types of traveller to visit, for example, Mallorca.

5 **Stagnation**: The number of visitors peaks and starts to go down. The resort has reached the carrying capacity and a number of conflicts and problems will overtake the desire to cater for more tourists. The resort is no longer as sustainable as it was and there are a number of economic, social and environmental problems. A surplus bed capacity starts to appear, for example, Costa del Sol.

6 **Decline or rejuvenation**: The tourist market becomes saturated to the point that a decision needs to be taken to re-invest in the area or to allow the area to continue to stagnate and decline. In decline, the tourist numbers will continue to drop and the resort will not be able to compete with the new, shiny resorts elsewhere. The use of the resort might convert to a more local focus and be used for weekends or day trips only. Tourist facilities will be unsustainable and will disappear, further plunging the resort into crisis, for example, Portrush.

Rejuvenation can save the popularity of a place; governments or travel companies might invest in new buildings, new leisure facilities and new attractions that change the focus for the resort, for example, Blackpool or spa towns in Europe.

Exam tip

Make sure that you know the specific details of each stage in the Butler model in as much detail as possible.

Summary

- The nature of tourism has changed a lot over the last 60 years. Mass tourism has developed into a global industry due to:
  - developments in transport
  - increases in disposable incomes
  - package holidays
  - internet access
- Tourism has brought some positive social and economic impacts.
- The demands of tourists and tourist resorts are changing over time. This can be seen through the pleasure periphery and the study of the Butler model.

Challenges and management of mass tourism

Challenges that arise from mass tourism

Pollution

The main challenge arising from any increase in mass tourism will be the increased environmental pressure that the area comes under. It is extremely difficult for tourism to be totally sustainable and often the **carrying capacity** will be quickly reached. So the more people that are coming to visit an area, the less able that area will be able to cope with the influx of people.

Pollution occurs when an area becomes damaged, contaminated or changed for the worse due to the actions of humankind. Pollution can come from a number of different sources (Table 17).

Carrying capacity is the maximum level that an area or resource can sustain without an unacceptable degree of deterioration of that resource.

Table 17 Challenges and consequences of pollution in tourist resorts

Type of pollution	Tourist activity	Challenges and consequences	Example
Water	Waste/sewage disposed of into rivers/lakes/sea from boats, ferries, cruise ships.	Build-up of waste which can contaminate water and kill animal/ plant life. Might stop water being used for tourism.	Mediterranean resorts, Florida
Atmosphere	Increased number of cheap flights and abundance of long-haul means more people flying than ever before.	Aviation oil burned high in the atmosphere can cause a lot of atmospheric pollution.	Flight paths and area surrounding Heathrow airport in London
Coasts/islands	Tourism resorts develop mostly along beaches/coastal seafronts and turn a natural landscape into an urban one. Roads and transport links are built.	Many natural coastal landscapes are lost (e.g. mangrove swamps, natural beaches along the south coast of England).	Mangroves in Belize; islands in Bali/ Indian Ocean are commercialised
Rural areas/ countryside	Tourists will go mountain climbing, trekking, walking, rambling, skiing and for picnics.	Infrastructure like ski huts and lifts needs to be built for mountain access. Users will cause pollution and damage to paths/trails and cause soil erosion.	Himalayas in Nepal; national parks in the UK
Vegetation	Trees will be cut down to clear areas for tourist resorts or to provide wood for buildings/ heat etc.	Deforestation and fires in parks can destroy natural habitats. Tourists coming into forest areas can change natural environments.	Amazon rainforest in Brazil
Wildlife	Unplanned hunting and fishing will have a negative impact on animal numbers and cause some to be endangered or even extinct. Safaris can unsettle animals and cause damage.	Feeding, breeding and migration patterns of animals can be upset causing a decline in numbers. Animals might become 'used' to humans. Hunting is still allowed in some places.	South Africa or Kenya: safari excursions or illegal hunting trips
Historic sites	Archaeological sites, historic sites or religious shrines might be visited by huge numbers of people.	Damage can be done by the visitors and can alter the appearance of a place.	Mecca in Saudi Arabia; Stonehenge in England
Urban areas/ tourist resorts	Hotels, restaurants, pubs and clubs and other entertainment facilities will greatly increase the urban footprint in a tourist area.	Increased building puts pressure on infrastructure and increased use of vehicular transport puts more air pollution and congestion in a place.	Palma in Mallorca; congestion in New York

Many of the key aspects of pollution in tourism can be found in relation to the protection and use of water. This annual mass movement of people puts huge pressure on the water services. Islands and resorts have to invest in elaborate plans to ensure there will be enough water for the peak season.

Overcrowding

As the huge numbers of international tourists continues to increase, overcrowding becomes an issue. Last year, some of the US Disney resorts in Disneyland and the Walt Disney World's Magic Kingdom had to start restricting access as they had reached their maximum capacity. The issue can be even more pressing in some major global cities. Some local people now refuse to go into certain parts of London due to the excessive number of tourists, and similar no-go areas are starting to appear at tourism hot spots in Istanbul, New York and Paris.

Honeypot sites

A honeypot site is a tourist resort that attracts a large number of tourists. This means that a lot of environmental and social pressure is put onto the local area and local people. Usually a honeypot will have one particular attribute that the surrounding area does not have; it might have attractive scenery or be a site of historical interest which will encourage large numbers of tourists to visit.

For example, although there were already many millions of tourists visiting Paris (46.7 million visited the area in 2015), the development of Disneyland Paris in Marne-la-Vallée on the outskirts of the city created another honeypot site. The theme park opened in 1992 and covers nearly 5,000 acres of land across the Disneyland Park, Walt Disney Studios Park, Golf Disneyland and the Disney Village. The park also includes seven Disney-themed hotels and six associated hotels in the local area. In 2015, 14.8 million people visited the resort, making it the top attraction in northern France.

Social sustainability

Social sustainability is the process that creates a sustainable, successful place where there is an emphasis on well-being, through the understanding of what people need from the places where they live and work. In the tourism context, it concentrates on how people of a particular place can retain their ethnic/social/cultural traits and differences, as they come into contact with people from a variety of different places. It is meant to create a better life for all people that also maintains a viable future. In relation to social sustainability in tourism, this means that there will be respect for human rights and equal opportunities for all.

Mass tourism often brings conflict to a place. If the expanding numbers and influence of tourism on an area gets out of control, it can negatively impact the social space and interaction with local people. Locals will look for different jobs in the service industry instead of traditional jobs. Young people will be particularly affected by any change due to tourism and will reject more traditional customs and roles in order to service the tourist needs.

Often 'leakage' can happen. This is when money generated from tourism is transferred back to another country, for example, in Kenya around 17% of tourism revenue escapes to MEDCs. This might be due to bookings being taken by large multi-national companies or through the purchase of goods from other countries. Ethically aware tourists are encouraged to support local businesses, to ensure that a fair amount of travel costs are payable to local suppliers, and local guides and support workers are employed in a way that educates, nurtures and facilitates investment in locals.

Some tourists fail to understand the cultural conflicts that can rage as they meet a local person. Sometimes the tourist might belittle the lifestyle, cultural heritage or life choices made by a particular group of people. They do not understand the wider social and cultural fabric of life in a particular area. They might not understand particular religious laws or social practices and fail to give the traditions and customs the respect that they are due.

Competition for resources

Any place where tourists and local people come into contact will cause a level of conflict that creates competition for resources. Land that might have been used for farming in the past will be used to build resorts. Conflicts take place over who gets preference for water supplies: drinking water for the local population or for swimming pools for the tourists.

Any development in tourism will require improvements and investment in the basic infrastructure — airports will need to be improved, transport links will need to be developed to move the travellers from airports to resorts. Tourist facilities will also need to be improved: accommodation, restaurants, bars, shops and entertainment all need to be provided to ensure that tourists are catered for, and also to ensure that every penny is squeezed out of them.

Local people will often feel that their needs are being overlooked in favour of the tourists. Living space will be taken and locals will feel that their society/cultural attributes are being set aside in order to make money from the tourists. It can be difficult for many locals to stand up to the will of big developers.

> **Exam tip**
>
> There are a number of challenges that can arise from the arrival of mass tourism in an area. Ensure you learn these and can quote relevant examples.

Evaluation of strategies to reduce the negative social and environmental impacts of mass tourism

The increase of mass tourism in any area will require careful management. A delicate balance needs to be maintained so that the needs of the tourists can be preserved — so that they have a good time and want to return — *as well as* the needs of local residents — so that they can gain employment and a purposeful existence that is not compromised by the influx of tourists but is further enhanced.

Reducing negative social impacts

We have already noted that conflict can exist between local residents and tourists. This means that a number of different strategies need to be used to manage the different aspirations and needs for each group so that order can be maintained. The effects on any host community can come from any indirect or direct interaction with tourists. Host communities are often the weaker party in any link as the guests and service providers try to leverage the best deal possible.

Changes can be brought to the local community: to family relationships, to traditional lifestyles, to shared ceremonies or moral outlooks. Tourism can destroy or corrupt indigenous culture. Tourism can destroy people's privacy, dignity and authenticity.

Table 18 Key positive and negative social impacts of tourism

Positives	Negatives
The influx of tourists will stimulate growth, new ideas and activities and develop businesses.	Local community will be lost within the increase of mass tourism. As resorts continue to expand and become city-resorts local traits and individuality will disappear as the resort becomes similar to other world resorts.
The customs and traditions of local people can be enhanced due to their interest to visitors; local cultural exhibitions, museums, food and crafts will be celebrated.	There will be major changes and a big decline in the celebration of local customs and traditions. Traditional ways of life will be 'modernised'.
Young people will have jobs and opportunities in the tourism industry which they might not have had in the past.	Property prices will be impacted by the increase in resort development. Local people might not be able to afford to live in the same area in which they grew up.
Employment opportunities are now more diverse. There is less emphasis on rural jobs as tourism develops.	Conflicts will develop as the residents feel that their needs are set aside in favour of those of the visitors.
	Any new infrastructure will cost a lot of money and will have minimal impact on helping to improve the lives of the locals. New hospitals might be too expensive for local people to use.

Planning policies need to be developed to ensure that further building is restricted and that infrastructure matches actual requirements. Local government policies need to be developed to protect the lives and lifestyles of local people. Many tourist resorts set up special committees to oversee any changes and to make sure they are sustainable.

Hamanasi (Belize)

Hamanasi is an adventure and dive resort to the south of Belize City in Belize. It is located on 30 acres of beachfront property which includes its own private nature reserve. It is described by Trip Advisor as being one of the top 25 hotels in the world for service. Accommodation consists of luxury treehouses in the forest canopy and spacious beach-front rooms.

The management at Hamanasi take their social impacts seriously. They say, 'We understand the importance of preserving our environment and indigenous communities.' They are active members of the local Hopkins Village and have taken a number of measures/strategies to support projects in the local community.

Reducing negative environmental impacts

Tourism can be a source of ecological destruction. It can be part of the solution to environmental problems (see next section on Ecotourism) but, more often, the pressures of mass tourism encourage fast growth of tourist facilities that convert rural areas into urban tourist resorts. Every aspect of the modern tourist resort will have an environmental downside. The building of one tourist hotel will cause a built-up footprint as a 5–6 floor structure is erected. A 500-room hotel will usually have over 600 flushing toilets, 500 showers/baths plus numerous swimming pools. The hotel laundry service might have to deal with over 4,000 towels per day, bed linen, table

> **Exam tip**
>
> When mass tourism is present, measures need to be taken to reduce the social and environmental impacts. Make sure you can evaluate these strategies in some depth, looking at both the positive and negative aspects.

linen and staff uniforms. Cleaning fluids will be used which could have an impact on the environment. Transport links including trains, buses, taxis and cars will all be required, bringing noise and air pollution in their wake.

Table 19 Key positive and negative environmental impacts of tourism

Positives	Negatives
Fragile environments might get more publicity which allows further protection and conservation practices for endangered wildlife and delicate landscapes.	Fragile environments will be damaged, e.g. footpaths and soil can be eroded. Mass tourism will increase the number of people visiting beaches, parks and natural areas.
There is a much better appreciation of the value of the landscape and cultural heritage.	Damage to local wildlife and landscape will come from overcrowding of the area. Concrete jungles will increase and the land is not able to recover.
Old buildings will be renovated and reused as part of the development process.	Increased transport will lead to congestion and more pollution in the local area.
	Different types of pollution will become more obvious: air, noise, visual and water pollution will all need to be monitored.
	Any increase in number of tourists will require further development of tourism facilities that will use up greenfield sites.

Hamanasi (Belize)

The Hamanasi Resort has long been noted for the approach that it takes in trying to help manage the local environment in a responsible manner. The resort prides itself in offering opportunities to visit local Mayan archaeological sites, have reef adventures — snorkelling or diving along the Belize Barrier Reef — or to explore the local tropical rainforest.

Case study

A tourism management policy at a national/regional scale

The Balearic Islands are one of the most popular tourist destinations in Spain. They are no more than four hours from most European countries. A record 9.6 million tourists arrived in Spain during July 2016 alone, out of an estimated 67 million tourists during the year. Around 13 million of these visitors were from the UK.

The influx of these tourists puts the 0.9 million residents under a lot of pressure, although 90% of the island's GDP is generated through tourism-related activities.

Ports in the Balearic Islands (like Palma de Mallorca) are being intensively visited by cruise ships; nearly 20,000 cruise-ship passengers arrive in Palma every Tuesday.

However, the tourists are not as welcome as they once were. In Palma, the capital of Mallorca, many anti-tourist slogans have been painted onto walls including 'tourists go home and 'tourist: you are the terrorist'.

Negative social impacts in Mallorca

The massive number of people arriving on the tiny island of Mallorca was always going to create a series of social issues. These include:

1 **Social behaviour**: The rise of mass tourism has encouraged groups of young people to travel and enjoy themselves, increasing alcohol and drug abuse, creating issues with prostitution and trafficking, and creating drinking cultures where English, Irish and German bars have supplanted traditional Spanish culture. →

2 Crime and security: Until the mass tourism influx, there was little crime on the islands. The huge number of visitors, intent on enjoying themselves, has led to serious behavioural issues on the island. The number of rapes and murders has risen steadily and the police have had to quickly learn how to mediate international law.

3 Language: Over time, the Spanish language has gradually been relegated to second place as visitors continue to demand the use of their home language. Entertainment establishments advertise in European languages and cater directly for particular nationalities to bring a taste from home.

4 Cultural conflict: Relationships between the residents and tourists often hang in a delicate balance. Locals resent the pressures on life that tourists bring, yet they often are employed in the tourist industry. Residents generally do not like to share facilities with tourists.

Negative impacts on Mallorca

In the summer months the population of Mallorca swells considerably. This can put a huge amount of pressure on the local environment in the following ways.

1 Landscape transformation: In the 1960s there was a massive building programme across the island which converted the coastal features and sand dunes into urban landscapes with tall apartment blocks and hotels. Resorts such as Calvia, Magaluf, Alcudia and Palma transformed the coastal landscapes. Animal habitats were destroyed and many species of animals disappeared from the island.

2 Water supplies: Access to a clean, reliable supply of drinking water has always been an issue. However, with the influx of tourism the problem has been further accentuated. Restrictions are applied on hosepipes and watering gardens. Residents get increasingly frustrated as they know that hotels are using more than their fair share of water for gardens, golf courses, swimming pools and showers. This adds to the conflict with tourists as a water-conscious farmer might use only 140 litres of water a day, a city resident might use 250 litres, while a mid-range tourist might use 440 litres

and a high-range tourist might use 880 litres. Water is an over-used resource on the island.

3 Sewage and waste: The amount of sewage and waste produced on the island is increasing at a fast rate. Sewage treatment plants are old and outdated and some still pump raw sewage directly into the Mediterranean, causing huge pollution issues for the local government. The Mallorcan government estimate that the typical tourist generates 50% more rubbish than a local resident.

Strategies to reduce the social impacts

1 Social behaviour: Increases in the number of tourists in 2015 and 2016 have caused local police to introduce new anti-social behaviour laws. In Playa de Palma, police regularly patrol the beaches to enforce a new rule that does not allow alcohol consumption in public places. Drinks are only permitted in bars and clubs. This move is aimed at changing the public perception of tourism in the area.

2 Crime and security: Recent increases in crime and anti-social behaviour in tourist resorts have meant that local authorities need to spend more money on police and security services. Spanish authorities want visitors to enjoy their time in the sun and so work closely with police forces across Europe. A new tourism tax was introduced from 1 July 2016 which charges around 2 euro per person per night and monies collected are earmarked for improvement measures.

3 Language: Local community and cultural groups have been lobbying for the Spanish and Catalan languages to be given more dominance in streets and transactions. New courses and classes are offered to tourists to help them understand the local culture.

4 Cultural conflict: Although the vast majority of local residents rely on tourists for their source of income, they are often frustrated by the behaviour and selfishness displayed by tourists. There are many schemes built up around resorts to protect and separate local residential and tourist areas. Local guides take cultural meetings and celebrations in hotels and often share the local traditions with their guests. Overcrowding of beaches and local facilities can →

bring conflict, so management strategies have been put in place to ensure that there is enough space for everyone.

Strategies to reduce the environmental impacts

1 **Landscape transformation**: Legislation helped to protect planning procedures and protect areas in four categories:
   - parks (national and nature parks)
   - nature reserves
   - natural monuments
   - protected landscapes

   In addition, the Coastal Law of 1988 guaranteed the protection of unspoiled areas and fragile ecosystems. The National Park of the Archipelago de Cabrera was created in 1991. New building on unspoiled land was severely curtailed.

2 **Water supply:** The regional Balearic Autonomous Community (BAC) government recently introduced a water management strategy that would:

   - introduce measures to make more effective use of the existing water supplies by reusing waste water for farming and golf course irrigation
   - repair distribution systems and reduce leakages
   - encourage conservation of water through water metering
   - develop new sources of supply through bore-holes and the expansion of desalination processes
   - artificially recharge aquifers by piping water from places that are in surplus

3 **Sewage and waste**: New measures have been put in place to try to manage sewage and waste. Recycling activities are actively encouraged on the island. In recent times Mallorca was the biggest incinerator of waste in southern Europe (destroying 84% of all waste in this manner). However, it aims to reduce this to 62% over the next 15 years.

Exam tip

Mallorca is a great example of what can happen when mass tourism hits an island. Make sure you can clearly identify the strategies/management policy used here and be able to evaluate the measures taken.

Summary

- A number of challenges have arisen from the development of mass tourism: pollution, overcrowding, honeypot sites, social sustainability, competition for resources.

- A range of strategies can be used in order to reduce the negative social and environmental impacts that mass tourism can bring. These strategies can bring both positive and negative consequences.

Ecotourism: opportunities, challenges and regulation

Definition of ecotourism

There has long been a debate about what exactly ecotourism is. The UNWTO definition of ecotourism notes that it is any tourism which has the following characteristics:

1 All nature-based forms of tourism in which the main motivation of the tourists is the observation and appreciation of nature as well as the traditional cultures prevailing in natural areas.

2 It contains educational and interpretation areas.

3 It is generally, but not exclusively, organised by specialised tour operators for small groups. Service provider partners at the destinations tend to be small, locally owned businesses.

4 It minimises negative impacts on the natural and socio-cultural environment.

5 It supports the maintenance of natural areas.

Figure 35 The three main pillars of ecotourism

Therefore, ecotourism encourages visitors to be responsible for leaving only a small carbon footprint, and to ensure that there is benefit for the local community and the environment. It is a type of sustainable development but its aim is to reduce any impact on the naturally beautiful environments.

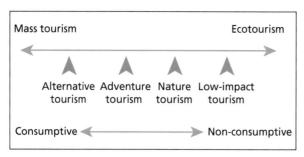

Figure 36 The tourism continuum

Exam tip

Make sure that you have a clear understanding of what ecotourism is.

Social, economic and environmental benefits of ecotourism

Knowledge check 25

What is ecotourism?

Social benefits

- **Cultural preservation** Visitors will also be able to learn and engage with local people, traditions and cultures. This allows the preservation and celebration of traditional culture, which might be otherwise lost. Traditional dress and music can be preserved as visitors will pay to see examples of cultural practices in situ.

- **Local arts and crafts** Travellers love to collect exotic examples of local art and craft work from around the world. The preservation of these crafts is important to the tribal/cultural identity of the local people.

- **Religion** In some places, religion has become an attraction as religious buildings and events have become important reasons to visit.

Economic benefits

- **Employment** An increasing number of job opportunities can be created due to ecotourism. Travellers are usually rich and are willing to pay a lot of money for luxury services. There are many jobs for the local people in resorts, as guides, working in building and servicing accommodation, making merchandise and providing food, drink and entertainment for guests.
- **Hard currency** As above, the rich tourists will bring a lot of hard currency (e.g. dollars, pounds and euros) which can be converted at higher exchange rates in some economies.
- **Reduction of poverty** Tourism, especially in LEDC countries, can bring jobs and money to people in rural environments who might not expect a chance of generating a good income. Workers are paid steady wages and can often benefit from extra health and social benefits.

Environmental benefits

- **Preservation** As people visit some of the most beautiful places on the planet, they will help to safeguard it. Nature and wildlife can be protected, offering a sustainable opportunity of employment for local people to help protect and conserve the environment, paid for by the visitor.
- **Ancient and historical sites** Many ancient sites including castles and places like the Taj Mahal in India can be protected through mechanisms such as the UNESCO World Heritage Site designation. Ecotourism can help protect these sites, allowing access and generating income but with careful management plans in operation.
- **Marine and coastal areas** Some marine animals and areas such as the Great Barrier Reef in Australia have allowed fragile environments to be protected.
- **Biodiversity and endangered species** Conservation areas can be used to protect fragile animals and their habitats. Protection is put in place to maintain stable numbers of animals, to protect them from hunters and to ensure that they are around for future generations.
- **Education** Ecotourism demands education for both the host communities and the traveller. All developers are expected to include educational programmes as part of any tourist activity.

Social, economic and environmental negative impacts of ecotourism

Social negatives

- **Displacement of local communities** The money that comes into a community as a result of ecotourism can create further conflict and change the land-use rights of the local people. They might lose access to traditional homelands as more land is needed to expand tourist complexes and resorts. Villages might have to be moved to facilitate creation of the 'best view' for the luxury traveller. Land can be taken by governments and given to development companies directly. This will restrict local farming practices. Equally, if safari lodges and hotels are using wells to take up the groundwater for the use of guests, there might not be any left for local people and they might have to relocate to find consistent water supplies.

- **Threats to indigenous cultures** Although an ecotourist might have the intention of witnessing authenticity, the reality might be that ritual materials and heirlooms are sold and lost forever.
- **Tradition** The dignity and respect of the local population might be compromised as they change and adapt their culture to entertain the visitor.
- **Language** Many visitors expect their hosts to speak 'their' language and often English is considered as the language of tourism.
- **Religion** Conflict can occur as tourists come into contact with worshippers. Sometimes traditional religious sites and burial grounds are disturbed and built upon for tourism. Some religious practices can also go against religious law (e.g. sunbathing in Muslim countries).

Economic negatives

- **Leakage** This is when money that is generated from tourism is transferred back to another country, e.g. in Kenya around 17% of tourism revenue will escape back to MEDCs. This might be due to bookings being taken by large multi-national companies or through the purchase of goods from other countries. Ethically aware tourists are encouraged to support local businesses, to ensure that a fair amount of travel costs are payable to local suppliers and local guides and support workers are employed in a way that educates, nurtures and facilitates investment in the local worker.

Environmental negatives

- **Greenwashing** This term has been used as a form of spin/PR to market the idea that a place or organisation is more environmentally friendly or sensitive than what it really is. This is extremely common in relation to ecotourism. The term was originally used to describe the hotel industry's attempts to amplify their green credentials by telling guests to reuse towels to save the environment. Jay Westervelt noted that most hotels had no care for the environment but were more interested in making further profit.
- **Damage to fragile environments** Any tourism can have a detrimental impact on the surroundings. The transport we use, the food and drink we consume or the source of water for washing can all have serious implications for nature.
- **Water resources** Water is a scare commodity in many locations. Resorts might struggle to get access to steady water supplies. Sometimes wells and boreholes will be used and these might take water away from local people. In places where swimming pools are built, excessive amounts of water are required.
- **Energy resources** Tourists today use and demand access to a lot of energy for charging mobile devices and technology. This often means that resorts have to find ways to generate this electricity and this sometimes does not come from a renewable source.
- **Waste** Tourists generate a lot of waste, both solid and liquid, and although attempts might be made to recycle or treat liquid waste, often this is not done.
- **Animals** The sale of some locally made products can involve the killing of protected animal species (e.g. sea turtles, reptiles, wild birds or elephant tusks).

Case study

The benefits and negative impacts on small-scale ecotourism

The Monteverde Cloud Forest is a biological reserve stretching over 26,000 acres in Costa Rica with over 3,000 species of plants, 755 different species of tree, 100 mammals, 400 species of bird and 120 reptiles. It is a national reserve and is run by the Tropical Science Center (TSC), a non-governmental scientific and environmental organisation that was set up in 1962. Each year over 70,000 people visit the forest. It has quickly become a world leader in sustainable tourism.

The Arco Iris Lodge is an ecotourism destination in the cloud forest. Located on the outskirts of the Monteverde reserve, accommodation is provided in specially designed cabins in preserved forest. The area is famous for its high concentration of exotic birds. Some travel writers think the area is a model for ecotourism.

Benefits of ecotourism in Monteverde

Social benefits

- Incomes in the area promote further developments in conservation and community projects. These include comprehensive local environmental education and maintenance programmes, e.g. one programme has been ongoing for over 20 years where thousands of local students have been given classes on cloud forest ecology and environmental issues.

Economic benefits

- Ecotourism has brought a lucrative source of income to the people living in the area. The Tropical Science Center is funded through the tourism activities. In Costa Rica ecotourism brings in around $2billion a year. Every visitor to the cloud forest must pay a $15 entrance fee which is used to conserve the forest.
- The ecotourism project has brought over 70,000 visitors a year into the forest. This means that people have jobs providing services for the visitors as guides, in shops, restaurants and bars and in the eco-lodges. The cloud forest has over 600 direct and indirect employees. If an area can move towards an ecotourism economy

there is less incentive for people to farm or cut trees down for income. As the number of visitors increases more people are being employed.
- In Costa Rica, much of the ecotourism is kept at a local level which means that profits are retained within the country: 80% of income in 2015 remained in the area (i.e. very little leakage).

Environmental benefits

- The development of the reserve protects the area from logging, farming and deforestation. This protects the biodiversity of the area but still allows the local people to live in a sustainable manner.
- The area is carefully managed. Only 2% of the forest is accessible to visitors. The rest of the forest is part of an 'absolute protection zone' where the flora and fauna are allowed to grow naturally. The creation of this area has protected the forest from major logging developments and commercial logging is not permitted.
- A number of eco-lodges and hotels have been developed inside the cloud forest. Some of these were originally made from lodges left over by loggers. Some of the lodges are basic but others offer a more luxurious experience in the high canopy of the rainforest. The lodges have been built to high environmental standards (gaining a Certificate for Sustainable Tourism).
- Some of the practices for environmental sustainability in the Monteverde reserve eco-lodges include:
  - not keeping animals in captivity
  - using biodegradable products in the kitchen, laundry room and cosmetics
  - treating black and grey waters through water treatment plants

Negative impacts of ecotourism in Monteverde

Social negatives

- Originally only small numbers of visitors would visit the cloud forest. However, there has recently been an influx of visitors which can put more pressure on the services provided.

→

- Many of the traditional ways of life and working have been lost as people now work in the service industries to provide for the needs of the visitors. Local people might find that they have to hold a few jobs as they might be low-paid or seasonal.
- Increased numbers of visitors can also bring conflict and overcapacity. Tourists might come under the influence of alcohol or drugs and there might be an increase in the amount of anti-social behaviour or crime. There has also been an increase in the number of people who have become sex workers and incidences of sexually transmitted disease have also increased, putting further pressure on local health services.

Economic negatives

- Ecotourism often quickly becomes big business and large global tourism companies can get involved so that any profits made will leak out of the country back to head office.
- Land prices in the local area have increased which makes it increasingly difficult for local young people to build houses so that they can stay living in the area.
- Increased tourism has also meant that more people have migrated into the area to live and work or look for jobs in the tourist industry facilities.
- There is also a real danger for overcapacity in the area. As the number of tourists increases, many people fear that this will stretch local resources and services to the limit, without any noticeable benefits to the local economy.

Environmental negatives

- Restricting access to a large area like this can be difficult to manage. Protecting the areas can be difficult, especially if there are not enough staff. It is possible for areas of the forest to be lost to developers and farmers before any action can be taken.
- The population of the area has started to grow, especially on the edges of the cloud forest and the local town of Monteverde. This puts increasing pressure on the local ecosystem, causes deforestation and loss of animal habitat and sometimes there is fragmentation of the forest.
- The increase in ecotourism, while bringing much-needed cash into the area, also puts pressure on the local ecosystem. Over 250,000 people visited Monteverde in 2015 which required additional building of accommodation, facilities and transport.
- The additional facilities required for tourists means that more electricity needs to be produced from sustainable sources, more fossil fuels are required for taxis and transport and more water resources are required for the hotel services. Getting rid of waste from hotels has also become an issue.
- Even the most 'eco-friendly' tourist lodges will have a negative environmental footprint. Increases in visitor numbers have led to a rise in the number of trails and tracks developed through the forest and the building of hanging bridges and high ropes courses through the upper canopy of the cloud forest.

Challenges in establishing effective international regulation of ecotourism

There has been a call from many environmentalists for one single global accreditation standard that could be applied to governments, hotels, tour operators, travel agents, guides, airlines, local authorities and conservation organisations. Many individual countries do operate their own regulatory programmes, for example, in Costa Rica the Certificate of Sustainable Tourism (CST) programme is designed to help balance the effect that business has on the local environment. Ecotourism Kenya is a similar programme that promotes responsible tourism practices. It features an eco-rating certification scheme for tourist accommodation facilities. Currently there are 31 facilities in receipt of the Gold certification, 57 for Silver and 23 at the Bronze level.

Exam tip

There is a long list of measures used to try to regulate ecotourism. See if you can find out more detail about each one.

The Quebec Declaration

2002 was designated International Year of Ecotourism (IYE). The World Ecotourism summit was held in Quebec City, Canada on 19–22 May. This was set up to be the main event for the IYE.

This was the first ever global ecotourism summit under the leadership of the United Nations Environment Programme (UNEP) and the United Nations World Tourism Organization (UNWTO). The final declaration was not a negotiated document, but a set of recommendations for the development of ecotourism activities in the context of sustainable development.

Among other things, it would:

- recognise that ecotourism embraces the principles of sustainable tourism, concerning the economic, social and environmental impacts of tourism
- acknowledge that tourism has significant and complex social, economic and environmental implications, which can bring both benefits and costs to the environment and local communities
- recognise that ecotourism has provided a leadership role in introducing sustainability practices to the tourism sector
- emphasise that ecotourism should continue to contribute to make the overall tourism industry more sustainable, by increasing economic and social benefits for host communities, actively contributing to the conservation of natural resources and the cultural integrity of host communities, and by increasing awareness of all travellers towards the conservation of natural and cultural heritage
- stress that to achieve equitable social, economic and environmental benefits from ecotourism and other forms of tourism in natural areas, and to minimise or avoid potential negative impacts, participative planning mechanisms are needed that allow local and indigenous communities, in a transparent way, to define and regulate the use of their areas at the local level, including the right to opt out of tourism development

Global Ecotourism Conference, 2007

The Global Ecotourism conference 2007 (GEC07) was held in Oslo, Norway on 14–16 May and represented the fifth anniversary of the Quebec Summit. Its main objectives were to assess the achievements and challenges that had hit ecotourism since 2002 and to bring further commitment and action to strengthen the contribution that ecotourism makes to conservation and sustainable development.

The conference was organised by The International Ecotourism Society (TIES) and the United National Environment Programme (UNEP). In all, 450 delegates from over 70 countries reviewed the Quebec declaration to evaluate how tourism could be made more sustainable.

A number of challenges remained in ecotourism, such as:

- Interest in visiting natural areas, experiencing authentic local living, and observing wildlife has continued to grow, bringing opportunities but also pressures, and the increasing need for best practice management.
- The term ecotourism is more widely recognised and used, but it is also abused, as it is not sufficiently anchored to the definition. The ecotourism community, therefore, continues to face significant challenges in awareness building and education and actively working against greenwashing in the tourism industry.

> **Knowledge check 26**
>
> Why do some governments feel that there needs to be regulation to protect places of value?

- Stronger leadership and strategies are needed in order to substantially decrease ecotourism's carbon footprint generated from multiple sources including facility operations and transport-related greenhouse gas emissions.

Some of the main recommendations from the conference were that international agencies, governments and those involved in planning and delivering ecotourism were called to:

- recognise the valuable role that ecotourism plays in local sustainable development
- maximise the potential of well-managed ecotourism as a key economic force for the conservation of tangible and intangible natural and cultural heritage
- support the viability and performance of ecotourism enterprises and activities through effective marketing, education and training
- address some of the critical issues facing ecotourism in strengthening its sustainability

Green Globe scheme

The Green Globe certification scheme is an assessment scheme that looks at the sustainability performance of travel businesses. Through this the businesses can monitor improvements and can achieve the Green Globe Standard which includes 44 core criteria using 380 compliance indicators.

The core criteria are broken down into four categories:

1 Sustainable management
2 Social/Economic
3 Cultural heritage
4 Environmental

There are three levels of recognition:

1 **Green Globe Certified Member**: Awarded to members that are certified against all of the criteria in the Green Globe Standard. Certification. It is confirmed annually when more than 50% of the criteria have been achieved.

2 **Green Globe Gold Member**: This is awarded to Green Globe members that are certified for five consecutive years. It is not automatically granted but after a review of overall performance during the five years to ensure that continuous improvement has been made.

3 **Green Globe Platinum Member**: This is awarded to Green Globe members that are certified for ten consecutive years. It is not automatically granted but after a review of overall performance during the ten years to ensure that continuous improvement has been made.

Green Globe Lite is an online version of the programme, which allows companies to achieve the specific Earth Check certification using online tools.

UNESCO World Heritage Sites

A World Heritage Site is a landmark that has been officially designated by the United Nations Educational, Scientific and Cultural Organization (UNESCO). The sites are noted as having some form of cultural, historical or scientific value or significance and they are legally protected by international treaty. The sites are seen as having extreme importance to humanity.

The programme started on 16 November 1972 and in January 2017 there were 1,052 sites — 814 cultural, 203 natural and 35 mixed properties — across 165 countries.

UNESCO notes their strategic objectives (the 5Cs) as:

1 Strengthen the **credibility** of the World Heritage list
2 Ensure the effective **conservation** of World Heritage properties
3 Promote the development of **capacity-building** measures
4 Increase public awareness, involvement and support for World Heritage through **communication**
5 Enhance the role of **communities** in the implementation of the World Heritage Convention

To be included on the list, sites need to have an aspect of 'outstanding universal value' and must meet one out of ten selection criteria.

In order for an area to be nominated, the country must complete an inventory of its significant ethical and natural properties (called the Tentative List). Any property named on this list can then be placed into a Nomination File. This is evaluated by the International Council on Monuments and Sites at the World Conservation Union. Recommendations are then made to the World Heritage Committee, which meets once a year to make decisions about inscription.

The Operational Guideline for World Heritage Sites notes that the protection and management of World Heritage properties 'should ensure their Outstanding Universal Value . . . there must be adequate long-term legislative, regulatory, institutional and/or traditional protection and management to ensure their safeguarding'.

Each property should have a management plan that will specify how the area is to be preserved and protected for present and future generations.

Summary

- Ecotourism has become a big element of the tourism market in recent years.
- Ecotourism can bring some social, economic and environmental benefits. However, it can also bring some negative impacts:
  - social (displacement of local communities and threats to indigenous cultures)
  - economic (leakage)
  - environmental (greenwashing and damage to fragile environments).
- Challenges in establishing effective international regulation for ecotourism have been addressed through the following measures to regulate ecotourism:
  - the Quebec Declaration
  - Global Ecotourism Conference 2007
  - Green Globe Scheme
  - UNESCO World Heritage Sites

Questions & Answers

The A2 Unit 2 Geography paper contains two questions for each of the four options. **Students answer two questions — one from each of their two chosen options.**

Each question is awarded up to 35 marks, giving a total mark out of 70.

	Compulsory?	Marks (out of 70)	Exam timing (out of 90 minutes)
Option A Cultural geography			
Q1 and Q2	No	35	45
Option B Planning for sustainable settlements			
Q3 and Q4	No	35	45
Option C Ethnic diversity			
Q5 and Q6	No	35	45
Option D Tourism			
Q7 and Q8	No	35	45

Examination skills

As with all A-level exams there is little room for error if you want to get the best grade. Gaining a grade A* is not easy in A level geography so you need to ensure that every mark counts.

The examination papers for A2 Unit 1, Unit 2 and Unit 3 are all 1 hour 30 minutes long. There are 70 marks available for Unit 2, which means that you get just over 1 mark for every minute to work your way through the paper. You need to make sure that you manage your time carefully — you have 45 minutes to finish answering each of the two questions. If you find that you have time left over in this exam, the chances are that you have done something wrong.

Exam technique

Students often find it difficult to break an exam question down into its component parts. On CCEA exam papers, the questions are often long and difficult to understand, so you need to work out what the question is asking before you move forward. One difference between the AS and A2 examinations is that the questions for A2 geography are set in one paper and the resources are contained in a separate booklet. Candidates will need to make sure that they refer to these resources carefully. The resources could be text (from a range of sources), maps, diagrams or photographs. You need to ensure that you sort through the information carefully and use the information from the resource **to help you** answer the question.

Command words

To break down the question properly, get into the habit of reading the question at least *three* times. When you do this it is sometimes a good idea to put a circle round any command or key words that are being used in the question.

A common mistake is failing to understand the task being set by a question. There is a huge difference between an answer asking for a discussion and one asking for an evaluation.

Structure your answer carefully

Sometimes the longer questions on exam papers can prevent students from achieving maximum marks. Questions that are marked from 8 to 18 marks will be marked using three levels. Later in this section we will look at some questions and give more guidance about how you should structure your answers.

One simple approach to consider is drawing up a brief plan for your answer so that you know where it is going and how you will cover all of the main aspects of the question. For example, you could draw a box to illustrate each element needed in an answer and fill each one with facts and figures to support the answer, using the marking guidance to help you work out how much time to spend on each section.

Show your depth of knowledge of a particular place/ case study

The extended writing questions on the exam paper are usually focused on giving the student the opportunity to apply knowledge and understanding of case study material to a particular question. It is really important to show what you know here.

Examiners are looking for specific and appropriate details, facts and figures to support your case. The better you know and understand your case studies, the higher the marks you can potentially achieve.

About this section

A practice test paper with exemplar answers is provided. This will help you to understand how to construct your answers in order to achieve the highest possible marks.

Some questions are followed by brief guidance on how to approach the question (shown by the icon ⓔ). Student responses are followed by comments indicating where credit is due. These are preceded by the icon ⓔ. In the weaker answers, they also point out areas for improvement, specific problems, and common errors such as lack of clarity, weak or non-existent development, irrelevance, misinterpretation of the question and mistaken meanings of terms.

■ Option A

Question 1 Cultural geography

(a) Explain why the global growth of cyberspace has been uneven.　　　　　(8 marks)

ⓔ Marks are awarded for an answer that explains some of the contrasts across the world that deal with economic, social and political issues. Answers will deal with at least two of the three issues.

Level 3 (6–8 marks): Answer addresses at least two of the issues and focuses on 'why' the differences have occurred rather than just a description of the patterns.

Level 2 (3–5 marks): A good quality answer that might only look at one issue or might look at two with less depth.

Level 1 (1–2 marks): Answer is a description rather than an explanation.

Student answer

(a) The amount of internet access that people get around the world is varied. People who live in rich countries such as the USA and the UK have high amounts of access. Over 80% of people in the UK have access to the internet. This is because people who live here can afford the expense of having broadband access or 4G access on their mobile phones. It is much easier to get access in the UK as there is more competition for business and competitive prices.

ⓔ **4/8 marks awarded** The answer has some detail about the global contrast but it only deals with some of the economic impacts. This restricts the answer to Level 2; the answer would need to address either more social or more political issues to go into Level 3.

(b) Examine the relationship between social inequalities and religion. You should refer to places that you have studied to illustrate your answer.　　　　　(9 marks)

ⓔ Many people feel that they experience social inequalities such as social exclusion and discrimination based on a range of issues. The answer here will need to refer to social exclusion and discrimination but should also look at how some people's religious beliefs play a role in how they get on with and integrate with others.

Level 3 (7–9 marks): There is good commentary on social inequalities (and possibly discrimination) through religion. There is some appropriate reference to places.

Level 2 (4–6 marks): A good quality answer that might only take a superficial look at the social inequalities through religion. Perhaps less detail in relation to places.

Level 1 (1–3 marks): The answer lacks understanding and does not address the key ideas behind the question.

(b) Across the world many people will experience social inequalities and discrimination because they follow a different religious belief compared with the majority population. This can mean that people are forced to live in separate enclaves within a city so that they can protect themselves and live in groups for safety (enclaves). Their rights might be compromised because of these beliefs. In some countries, Christians are not allowed to practise their beliefs or to bring a Bible into the country. If they are found out they will be persecuted and punished. They also might face huge amounts of discrimination; they might not be allowed to build religious buildings (like a mosque) or be allowed to stop work to pray the number of times that they usually have to. In France and Switzerland recently there have been laws passed that make it illegal for Muslim women to wear full-length burqas. In France, it has been illegal to wear any religious symbols like Christian crosses or burqas in schools since 2004.

ⓔ **7/9 marks awarded** This is a good description of how religious intolerance can impact social inequality and cause discrimination. There is some description and reference to places in the answer but the answer could have gone a little deeper to make sure that more examples were covered.

(c) Discuss the impacts of migration from both your small-scale case studies of out-migration and in-migration on:
- economic activity
- service provision

(18 marks)

ⓔ The answer here needs to focus on the impacts/implications of the migration movements in relation to both economic activity and service provision. Candidates will need to add detail about *both* their case study for out-migration and in-migration.

Level 3 (13–18 marks): The answer will include good detail that shows that the candidate has a command of the material. There is good detail in relation to both of the case studies *and* reference is made to both economic activity and service provision.

Level 2 (7–12 marks): At least three of the four tasks have been attempted in some detail including details on out-migration and in-migration. There might be a requirement for more depth and information.

Level 1 (1–6 marks): If more than one of the tasks is missing. The answer lacks understanding and does not address the key ideas behind the question. The answer might only address *one* case study or might only explain the impacts of either economic activity or service provision.

(c) In 1841 the population of the Blasket Islands was 153 and there were 28 households. The official census of the time shows that the population dropped by 30% between 1841 and 1851 from 153 to 97. This is most likely due to the potato famine which had hit Ireland during this time. However, the decrease was much worse for other surrounding Islands. By 1953 there were only 22 people left on the Island.

This out-migration of people from the Blasket Islands has had a major impact on the economic activity of the Island. The islanders were subsistence farmers who sold things such as rabbits, fish, lobsters, birds and eggs. There was some pastoral and arable farming but life was never easy on the island and there was never a surplus of cash from the jobs that they did. They were just about making enough to survive. Due to so many people leaving the Blasket Islands, many of the Islanders received remittances from family members who had migrated and some received passage to join their relatives in the USA. Another implication on economic activity was the fact that there was a big loss of the youthful population and therefore the economic duties of collecting bird eggs, some aspects of farming and rowing to and from the mainland, which were dangerous, became less viable as the workforce became older. By the 1950s, there were too few able-bodied men to row the boats and this was the final nail in the coffin for the islands.

Out-migration also had implications for service provision on the islands. There were very few services on the islands for people, for example, there was no church, no priest, no pub, no school (until 1860), no post office or telephone until 1930, no electricity, no sewage, no mains water and poor transport links. It was the impact of electricity that led to the evacuation of the islands by the government as it was concerned with the cost of taking electricity to the islands in the 1950s. The fewer the people who stayed, the fewer services there could be, and as the rest of the country modernised the young people did not want to stay and followed their dreams to the mainland and to the USA.

Many Turkish workers made the decision to travel to Munich in Germany and became guest workers. By 2006 there were 6.7 million foreigners living in Germany and 1.7 million of these held Turkish citizenship. In 2010, it was estimated that there were around 4 million Turks in Germany. Around 15% of the population of Munich is Turkish.

This immigration into Munich has had major impacts on the economic activity of the city. After the war, there was a serious labour shortage in Munich. In 1961, with the building of the Berlin Wall, many workers were required. There became a rise in the school leaving age and earlier retirement ages which therefore meant that there was a smaller workforce which is why it was important to bring workers over from other countries. However, it wasn't all positive impacts on the economic activity. Even in 1961 there were still Germans who remained unemployed and many felt the call for migrant workers was due to pressure from the USA to help stabilise Turkey. In 1967 and 1990, recession led to unemployment with many people in Munich losing their jobs while the Turks were able to remain in their low-paid jobs. This fuelled social and ethnic conflict in some cities. Also, much of the money that was earned by the Turks was sent back (remittances) to Turkey and out of the German economy. Around €10 billion a year leaves Germany's economy through remittances.

The in-migration on Munich has also had implications on the service provision. The Turks took the jobs that Germans didn't want to do and many of these were providing low cost personal services for the German population, e.g. driving trams and buses and also cleaning. There has also been a demand for new services such as food services, textiles, religious services etc., for the migrants. An increased amount of money has been spent on translation services as more leaflets were needed in Turkish and more students in schools where German isn't their first language. Another implication is the fact that it puts pressure on health care, schools and transport due to their being an increase number of people living in Munich and many of them are migrants.

ⓔ **14/18 marks awarded** This student has identified material that answers all four of the aspects in the question. However, the student could have gone into a little more detail about how these impacted the two particular places. Some good depth of understanding of each of the locations shown.

■ Option B

Question 2 Planning for sustainable settlements

(a) Explain how the urban ecological and carbon footprints can help us to understand sustainability.   (8 marks)

ⓔ Marks are awarded for an answer that describes the differences between the urban ecological footprint and carbon footprints. Answers should be balanced between the two.

Level 3 (6–8 marks): The answer addresses both of the footprints in some depth. There is a solid understanding of the subtle differences between the two ideas and their contribution towards an understanding of sustainability.

Level 2 (3–5 marks): A good quality answer that might only look at one of the footprints or might look at two with less depth.

Level 1 (1–2 marks): The answer is a description rather than an explanation.

Student answer

(a) Urban ecological footprints help us to measure the amount of environmental and ecological assets that the population of a city needs to produce the natural resources that it consumes. The footprint therefore is the amount of land required to produce the resources required for one person to live, for example, in England, the city of Winchester came out on top and had a footprint of 6.52 global hectares of land required for each citizen.

A carbon footprint looks at the total amount of greenhouse gas emissions that are created in an area. Both of these are useful when trying to look at the sustainability of an area. The higher the footprint, the less sustainable that a place is likely to be.

ⓔ **3/8 marks awarded** This answer does not fully address the question. The student starts with some good explanation of the urban ecological footprint but the contrast with the carbon footprint is weak. In addition, there is little discussion on how these two measures help us come to an understanding of sustainability.

(b) Study Figure 13 on page 33 which shows some of the issues and further challenges in urban water supply across the world (from the SWITCH report). Use the resource **to help you** explain how water supply can be managed sustainably in a city.

(9 marks)

ⓔ The resource comes from the SWITCH report that looked at the sustainability of water management. Students will be able to use the diagram to help them to describe some of the key issues that are linked to the sustainable management of water in a city. Students can use information from their case study of a city to help them add additional material to their answer.

Level 3 (7–9 marks): There is good detailed information taken from the resource and from the student's own additional material about how water supplies might be managed in a sustainable manner.

Level 2 (4–6 marks): A good quality answer that might only take a superficial look at the resource or that might ignore the advice to include further information.

Level 1 (1–3 marks): The answer lacks understanding and does not address the key ideas behind the question.

(b) The resource clearly shows that there are a broad range of challenges that can be noted when trying to ensure a clean, reliable and safe water supply for a city. As the population continues to grow and the city expands, there is going to be an increased requirement for further sustainable water supplies. This will put more pressure on the water sources and new sources of water might need to be explored. There needs to be a good deal of investment in the infrastructure and pipe systems. It is important that there are as few leaks as possible as this reduces the amount of water in the system and can also allow contamination into the city. Perhaps there are opportunities for new emerging technologies to be used that would measure and charge water use accordingly so that the heavier users will be able to pay for their fair share. It is important that government policies are written to ensure that water supply is sustainable and is planned for the future and that all measures will be able to meet demand in the future. In Belfast, the Belfast Agenda is one of a number of planning documents that investigates the impact that further development of Belfast will have on its public services. As well as managing the waste created in Belfast in a sustainable manner, it is important that water supplies and sewage networks are carefully managed. In recent times, NI water has carried out £160 million projects to upgrade the sewage system in Belfast and to reduce the risk of flooding and improve water quality. Further developments have also meant £90 million improvements in water mains pipes so that leaks would be reduced and the quality of drinking water would be maintained at a high level. Water audits, water monitoring and leak detection systems now help to ensure that less water is lost through the system.

e **8/9 marks awarded** The answer is not perfect and could have emphasised the sustainability of the projects in more depth but there is a good attempt to use the resource in the question and to link this to some of the facts and figures linked with a water management strategy in a city.

(c) **With reference to your case study of a city, evaluate the following traffic management strategies:**
- **integrated transport networks**
- **car parking**
- **cycling policies** (18 marks)

e The answer here needs to evaluate both the positives *and* the negatives of each of the three management strategies. Students need to make sure that they get an element of balance between each of the strategies.

Level 3 (13–18 marks): The answer will include good detail that shows that the candidate has a command of the material. There is good detail in relation to both of the case studies *and* reference is made to each of the management strategies listed in the question.

Level 2 (7–12 marks): At least two of the three strategies have been attempted in some detail including details related to the case study city. There might be a requirement for more depth and information.

Level 1 (1–6 marks): If more than one of the tasks is missing. The answer lacks understanding and does not address the key ideas behind the question. The answer might only address *one* of the strategies or might have limited amounts of case study material present.

(c) Freiburg is a city in Germany that has had a long proud tradition of trying to become a more sustainable city. It has developed a special policy where there are five pillars of their traffic policy that look at 1) extending their public transport network, 2) promoting cycling, 3) promoting pedestrian traffic, 4) promoting liveable streets and 5) limiting individualised motorised traffic.

1 Integrated transport networks: public transport in the city is successful and has been made into a highly successful integrated transport network. People use electric trams to move around the city. These are environmentally friendly as they run on sustainable electricity and do not have very big carbon emissions. There are over 73 million annual journeys on the trams. Travellers can then pick up bikes or switch to pedestrian transport when they reach tram stations. However, the trams are often cramped and locals complain that there are not enough trams available at peak times. There are often no bikes available for hire and people have to walk or hire taxis for part of the journey.

2 Cycling policies: over 27% of the people use bicycles as their main mode of transport around the city. There are now over 400 miles of cycle path around the city and it is easier for people to cycle in and out of the city than to take any other mode of transport. However, this is not a popular option on

days with bad weather or when people need to get groceries or to transport large items back to their houses. It can be a real restriction on what people can buy. Bike hire is popular but most people who ride a bike to work can get a large percentage of this paid for as part of a 'buy-a bike for work' scheme.

3 Car parking: people in Freiburg do not use cars often which makes things sustainable.

e **11/18 marks awarded** There is some good discussion about the first sustainable strategy — integrated transport networks; there is a balance between the good points and bad points but the second discussion on cycling is less detailed and the third strategy is limited. There needs to be balance between the three strategies and each needs more specific case study material.

▪Option C

Question 3 Ethnic diversity

(a) Identify and briefly describe the role that *any two* factors play in any attempt to define ethnicity.

(8 marks)

e This is a straightforward task; the specification outlines four main factors that are used to define ethnicity: race, nationality, language and religion.

Level 3 (6–8 marks): The answer addresses both factors in good depth (4 + 4 marks) that clearly show the role that the factor plays within any definition of ethnicity.

Level 2 (3–5 marks): A good quality answer that might only look at one issue or might look at two with less depth.

Level 1 (1–2 marks): The answer lacks detail and is filled with vague comments.

Student answer

(a) Getting one true definition of ethnicity is never easy. The idea of ethnicity is the process of belonging to a particular ethnic group. Some of the main differences between people that allow them to be members of different ethnic groups are race, nationality, language and religion. Race is an obvious factor that divides one person from another. There are three main racial groups around the world: Caucasian (white), Mongoloid (Asian) and Negroid (black). There are obvious differences in the skin tone and sometimes other physical features such as noses, hair and eyes which allow people to look different. Another factor is language. Over the world people speak lots of different languages so people do not always understand each other. Some people can speak lots of languages but most people only learn one.

e **4/8 marks awarded** This makes reference to two of the factors. However, the answer on race is much better than the answer on religion. A lot more detail and information is required for a higher mark.

(b) Study Figure 23 on page 56 which shows European migration to the USA. Use the resource to help you show understanding of the impact of international migration on creating ethnic diversity. *(9 marks)*

ⓔ The resource shows details of the amount of European emigration to the USA between 1820 and 1920. There are a number of figures and details that students should use in their reference to how migration from Europe will have made an impact on creating the diversity of the USA. In addition, students need to bring their own additional material to show understanding of the impact of international migration.

Level 3 (7–9 marks): A thorough answer that makes good use of the resource material and has brought additional material into the discussion. The extra material helps with the argument and helps to discuss how international migration can help create ethnic diversity.

Level 2 (4–6 marks): A good quality answer that will still have some good material, but the amount of depth and detail will be reduced. There might be inadequate use of the resource or less detail in the additional material.

Level 1 (1–3 marks): The answer lacks understanding and does not address the key ideas behind the question.

> **(b)** The resource shows that there were a lot of people who migrated from western Europe to the USA between 1820 and 1920. Over 4.4 million people migrated from Ireland, 5.5 million from Germany, 530,000 from France and 320,000 from Turkey. As each of these groups of people migrated into the USA they will have helped to change the diversity of the US accordingly. There is a wide range of people who have travelled to the US from across Europe. The UK is another country that has experienced a lot of international migration. The British Nationality. Act meant that people have moved into the UK from all over the world but from the British Empire in particular. Large numbers of people moved from India, Pakistan, South Africa and streams of migration developed between Canada, Australia and New Zealand. More recently, huge numbers of people from eastern Europe have moved into the UK which has further increased the ethnic diversity in the UK.

ⓔ **7/9 marks awarded** The answer has used the resource and has quoted figures to show an understanding of how people from lots of different countries have been able to create diversity in the USA. The answer also goes on to bring in additional material from the UK about numbers of people who are coming into the UK. There is some good detail but further facts would have brought additional marks.

(c) With reference to your case study of an ethnically diverse city, discuss the processes that maintain ethnic diversity. *(18 marks)*

ⓔ Emphasis needs to be on the urban case study. Students need to describe the different processes that were used to maintain ethnic diversity in the city. The main emphasis will be on segregation and multiculturalism.

Level 3 (13–18 marks): The answer will include good detail that shows that the candidate has a command of the material. There is good detail in relation to the case study city. There will be a balance when explaining the roles that both segregation and multiculturalism play.

Level 2 (7–12 marks): There might be an imbalance in the response between the two processes or there might be a requirement for more depth and information.

Level 1 (1–6 marks): The answer lacks understanding and does not address the key ideas behind the question. The answer might only address one of the two processes or might identify something else as a process rather than maintains diversity.

(c) In Belfast there is a high degree of segregation along ethnic lines. Many parts of the city are either Protestant or Catholic.

Segregation is when there is a separation of ethnic groups and the people live separate lives. This minimises opportunities for social interaction. In Belfast, there are high levels of religious segregation in the city. This is marked in many areas through the use of walls and murals which usually are used to project the local political thought and stories of cultural identity and celebrations of folk heroes and key moments. Marking of territory is supposed to strengthen community links within but also make any outsiders feel out of place, therefore in most areas there are only people with the same views, religious beliefs etc. Belfast is an extremely segregated city. The segregation was further increased in the early years of the 'Troubles'. In the 1960s, 67% lived in segregated streets and by the 1980s it had risen to 78%. This is because of the Troubles and people were intimidated out of their homes where they didn't feel safe and relocated into areas perceived as safer.

Multiculturalism is when there are different cultures in society and instead of assimilation, there is celebration of their cultures. This means that cultures remain unique within society. Belfast has been made up of competing Roman Catholic and Protestant communities for a long time. It has never really been multicultural in the racial sense but it definitely is in the religious sense and in the sense of how these religious groups have attached themselves to feelings of a perceived national identity. Throughout the Troubles, the culture of Belfast became evident. People of different religious groups respected their religious beliefs and expressed them through various forms of art, such as poems, songs, paintings and murals. Each religious group had their own distinctive style of each art form and often enforced the unity of a neighbourhood.

Immigrants who have settled in Belfast also get the opportunity to share their cultural traditions. As Chinese people settled in NI there are Dragon Boat races on the river Lagan and the Indian community share Mala and Diwali. There are also restaurants in the city for those who enjoy eating Chinese, Indian, Italian, Japanese, Polish and Spanish meals.

In Belfast 75% of people said they thought the presence of foreigners was good for the city, higher than the EU average of 73%. By contrast, only 72% of people in London agreed.

@ **13/18 marks awarded** There is some good depth in relation to each of the two processes that are identified as being important to the maintenance of ethnic diversity. Case study material shows knowledge of the city but the answer could have gone a little further in relation to each of the two processes to talk about some of the recent changes that have happened in the city, for example, more place names and examples of segregation in north and west Belfast would have added to the answer.

■ Option D

Question 4 Tourism

(a) With the aid of a diagram, describe how the Butler model can be used to demonstrate the changes that a tourist resort might experience over time. (8 marks)

@ This is a straightforward task; students need to draw and label the diagram carefully (worth up to 4 marks) as shown below:

- Award 1 mark for the correct labelling of both of the axes, then
- Award 3 marks for the labelling of all of the stages through the model, or
- Award 2 marks for the labelling of four of the stages correctly, or
- Award 1 mark for the correct identification of fewer than four stages.

The additional 4 marks can then be awarded for the discussion on how this model can be used to show the changes that a resort might experience over time.

Award 3–4 marks for a detailed description of the changes over time. The answer might refer to particular resorts but this is not required in the question. All stages of the model should be explained without exception.

Award 1–2 marks for a more basic description of the changes over time. The answer might not cover all of the stages in the model or might not go beyond a simple statement of the stages.

Student answer

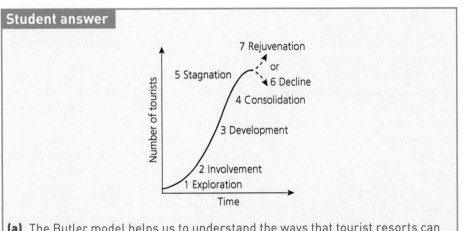

(a) The Butler model helps us to understand the ways that tourist resorts can change over time.

Exploration: the resort will not have many visitors at this point. It will just be a simple village with few facilities but people will still come to enjoy the simple pleasures it offers.

Involvement: the resort starts to grow and opens more accommodation for visitors and simple things for tourists to do. For example, in the 1960s resorts like Magaluf started to build apartments.

Development: the tourist numbers are now expanding towards mass tourism. Apartment blocks and hotels are springing up all over the place to capitalise on this.

Consolidation: tourism is becoming the most important business in the area and mass tourism has become the only game in town. Resorts like those in Mallorca are filled to capacity.

Stagnation: changing fashions in the travel business mean that people are trying other resorts. The older resorts are seen as old fashioned and they will either continue on their path to decline (like Portrush) or they will have to reinvest in their facilities and begin a process of rejuvenation.

e **8/8 marks awarded** The diagram is correct, the axes are labelled correctly and all of the stages in the model are positioned correctly on the diagram. The explanation of the model is detailed and has covered all of the stages appropriately.

(b) **Briefly explain some of the strategies that might be required to reduce the negative social and environmental impacts that mass tourism might bring.**   (9 marks)

e The specification asks the students to discuss some of strategies that could be used to reduce negative social and environmental impacts of mass tourism. There is no requirement in the question for places but students might make reference to one case study example; however, the emphasis in the answer needs to be on the description of the strategies.

Level 3 (7–9 marks): A thorough answer that has identified more than one strategy to reduce negative social impacts and more than one strategy to reduce negative environmental impacts.

Level 2 (4–6 marks): A good quality answer that will still have some good material, but the amount of depth and detail will be reduced. There might be an imbalance in the discussion of the social and environmental impacts.

Level 1 (1–3 marks): The answer lacks understanding and does not address the key ideas behind the question.

(b) Hamanasi is a special ecotourist resort in Belize. It is a good example of a place that is trying to make sure that the negative social and environmental impacts caused by mass tourism are avoided. The resort is keen to invite visitors to the country but wants to make sure that this does not cause problems for the people of Belize or the environment.

Social strategies: some of the social strategies will be there to make sure that the local people actually benefit from the tourism, rather than all of the money and benefits going to other people outside the country. Hamanasi will ensure that the vast majority of workers will come from the local area.

Profits from the resort are used to support local schools (the Backpack to School project) and to give regular health care to the local people.

Environmental strategies: when too many people come into an area this can bring environmental problems. The resort will try to limit the number of people who can actually come and will make sure that they build their houses using the local materials.

e **6/9 marks awarded** There is some good information in the answer about some strategies that might be used to reduce the social impacts but the environmental factors are not handled to the same depth. Therefore only a Level 2 can be award to this answer.

(c) **With reference to your small-scale case study, discuss some of the benefits and negatives that ecotourism can bring.** (18 marks)

e The specification clearly notes that the list of benefits can be either social, economic or environmental.

Students need to make sure that they include information about as many of the different benefits and negatives as possible. There needs to be good detail related to the place.

Level 3 (13–18 marks): The answer will include good detail that shows that the candidate has a command of the material. There is good detail in relation to the ecotourism case study. There will be a balance when explaining the different benefits and negatives and there will be some variety of explanation between social, economic and environmental issues.

Level 2 (7–12 marks): There might be an imbalance in the response between the benefits and negatives. There also might be an overemphasis on one aspect of the issues, for example, too much information about the environmental issues and not enough on the social and economic issues. There might be a requirement for more depth and information.

Level 1 (1–6 marks): The answer lacks understanding and does not address the key ideas behind the question. The answer might only address one of the two areas or the student might go off topic and not relate the information to an ecotourism resort.

(c) Kenya was one of the first countries in the world to embrace the growth of ecotourism.

Although tourism in Kenya is extremely important for income, providing jobs for local people etc., it can have many different consequences, particularly on the environment. For example, the minibus drivers in the Masai Mara who are hoping for tips often drive too close to the animals and this has caused them to change their natural behaviour patterns. Minibuses also churn up large areas of the bush by not staying on the roads. This is a negative environmental impact because it creates dust storms and soil erosion. Also, when several minibuses gather in one place, such as surrounding a pride of lions, this can frighten the animals. Balloon safaris have been criticised for frightening animals because the burners used to keep the balloons in the air are noisy and the shadows cast by the balloons can startle wildlife.

Another major environmental problem occurs when people walk on the coral and boats run aground or drag anchors on the reefs as these will damage parts of the fragile ecosystem. The shoreline itself has been degraded. The first hotels were built with little regard for the surrounding environment and do not blend in with the landscape but later developments have been more sensitive. However, it can still cause urbanisation due to the mass amounts of hotels, tented camps, accommodation blocks etc. The provision of swimming pools, clean water and electricity in hotels can cause resentment among local people who may have no running water or other basic utilities. Hotels use a massive amount of water due to swimming pools, laundry and cooking. This means that there is even less water available to local people who live within a ten-mile radius of the hotels. There can also be pollution caused by tourists such as sewage pollution which can be hard to handle, especially in a country like Kenya.

However, the money that the tourists bring into the area is useful for paying for the different conservation measures that are needed to keep the animals and ecosystems safe. Protected areas can be safeguarded and the different animals can be protected from poachers.

Foreign companies, which own 80% of the hotels and organise the package tours, supply much needed investment. However, they take most of the profits back home. Tourism provides jobs and a steady income for local people, improved quality of life, clinics/medical help, housing for workers etc., but the Kenyans are aware that they have to manage tourism effectively to ensure that the environment isn't damaged beyond repair.

℮ 11/18 marks awarded This answer has some good depth in relation to the ecotourism in Kenya. However, there are a number of problems with the answer. There is not really any good depth of knowledge in relation to the precise ecotourism case study; most of the information is done generally. In addition, the answer concentrates on the benefits and negatives of environmental impacts but does not really go into much detail in relation to social and economic impacts. There is some on each of these but a more detailed discussion is required.

Knowledge check answers

1 A cultural group is a community of people who share common experiences. The members of the group will have a similar outlook.

2 Cultural nationalism is a form of national sentiment/feeling/identity where the nation is defined by having a shared culture. The national identity of the place will be shaped by particular shared cultural traditions and language but not on ideas of a shared common ancestry or race.

3 Push factors are seen as the things that cause people to want to leave; these are the things that make them unhappy with their current life. Pull factors are the things that are attractive in the new place, the things that might be seen as being improved due to a move.

4 A series of barriers might get in the way and influence the decision to migrate; maybe a family member becomes sick, transportation costs increase or financial difficulties/recession make it hard to sell a house prior to leaving.

5 Undocumented migrants are people who do not have a permit of residence allowing them to stay in their country of destination. These are people who have not got a visa or permission to move into a country. They might have been unsuccessful in seeking asylum, overstayed their visa or might have entered the country illegally. Documented migrants are those who hold valid passports, visas and paperwork to allow them to travel legally from one country to another.

6 Migration has always been an issue in the UK; we have usually been a nation that sent lots of people to new places but the free movement of European citizens (documented migration), especially from the eastern European countries into the UK, brought this issue to the fore.

7 Cyberspace has become one of the most important developments over the last 20 years as people now have an information gateway that they can access through multiple devices. The way that people communicate and socialise with each other has been transformed.

8 Goal 11 — make cities inclusive, safe, resilient and sustainable — is the main SDG that focuses on sustainability in settlement.

9 The carbon footprint is the total set of greenhouse gas emissions caused by an individual, event, organisation or product expressed in its carbon dioxide equivalent. It is usually measured by the amount of carbon dioxide and methane emissions in a population. An urban ecological footprint is the amount of land required to produce the resources needed by one person (to support their lifestyle). It attempts to quantify the impact that one person can make on nature.

10 Water is one of the most important resources for life. Many see access to clean drinking water as a basic human right, yet in many places around the world it can be difficult to get access.

11 Eco-towns must achieve sustainability standards significantly above equivalent levels of development in existing towns and cities by setting out a range of challenging and stretching minimum standards.

12 The needs of the resident and the amount of money that the resident has to spend has changed considerably in Northern Ireland over the last 120 years. In the last few years the biggest growth has been in the subdivision of houses into flats for rental.

13 A brownfield site is a piece of land that has already been used and is now lying derelict. It is much more sustainable to use this land again rather than spoiling some green space at the edge of the city.

14 Sustainable transport can: allow for the basic needs of individuals, companies and society to be met safely; be affordable — operating fairly and effectively across a variety of transport modes; limit emissions and waste as much as possible; and use renewable resources where possible.

15 An ethnic group is a group of people whose members have a distinct culture which makes them different from the rest of the population. Membership of an ethnic group is usually characterised by a shared sense of heritage, origin, religion, art, physical appearance or ancestry. Ethnic groups are usually minority groups in the wider social and cultural context.

16 The main factors that influence ethnicity are the more obvious: race, nationality, language and religion. However, the perceptions of ethnic and social identity are not immediately as divisive or obvious. Also, there is more flexibility to move between some of these groups as they are less fixed.

17 Ethnic diversity is when there are multiple traditions/ethnic groups living in one particular area. This means that different ethnic groups with different racial characteristics, nationalities, language or religious beliefs will coexist. There will be ethnic mixing between the communities as they will come into contact with each other.

18 Migration is the single biggest reason why people from different ethnic communities have come into contact with each other. Migration is part of colonisation and annexation as both processes require a movement of people either to settle or to control the area.

19 Integration is the movement of minority groups into the mainstream society. Assimilation is the process where a person's culture comes to resemble that of another group. Full assimilation occurs when new members of a society are difficult to distinguish from the older members of the group.

20 Conflict is an inevitable outcome of ethnic diversity when the level of division has reached a certain point. Conflict can take a number of forms, from a strike or peaceful protest to a war. Conflict occurs

when people have a disagreement in respect to differences of interest, need or point of view.

21 Tourism is when people visit another region or area for a period of at least 24 hours. It can be international or within the country of the traveller. Mass tourism usually describes the practice where many tens of thousands of people go to the same resort at the same time of year. It is the most popular form of holiday as everything is packaged together so that the visitor can buy the accommodation, local transfers and flights together, saving money and making the cost of the holiday more affordable.

22 Social impacts will impact people and economic impacts will impact money and jobs.

23 The main idea behind this concept is that over time people have been able to travel further and further. The increases in prosperity and technology have enabled this movement, meaning that there are few places left on the planet that are not reachable within 24 hours of travel.

24 A honeypot site is a tourist resort that attracts a large number of tourists. This then means that a lot of environmental and social pressure will be put onto the local area and the local people. Usually a honeypot will have one particular attribute that the surrounding area does not have; it might have attractive scenery or perhaps a site of historical interest which will encourage large numbers of tourists to visit.

25 Ecotourism includes all nature-based forms of tourism in which the main motivation of the tourists is the observation and appreciation of nature as well as the traditional cultures prevailing in natural areas. It contains educational and interpretation areas. It is generally, but not exclusively, organised by specialised tour operators for small groups. Service provider partners at the destinations tend to be small locally owned businesses. It minimises negative impacts upon the natural and socio-cultural environment. It supports the maintenance of natural areas.

26 It is not enough to have special places around the world, we need to take action to ensure that they are around for many years to come. If organisations like the UN do not step in, then the areas will be exploited and lost.

Index